MATHEMATICAL THEORY OF WAVE MOTION

ELLIS HORWOOD SERIES IN MATHEMATICS AND ITS APPLICATIONS

Series Editor: Professor G. M. BELL, Chelsea College, University of London

The works in this series will survey recent research, and introduce new areas and up-to-date mathematical methods. Undergraduate texts on established topics will stimulate student interest by including present-day applications, and the series can also include selected volumes of lecture notes on important topics which need quick and early publication.

In all three ways it is hoped to render a valuable service to those who learn, teach, develop and use mathematics.

MATHEMATICAL THEORY OF WAVE MOTION
G. R. BALDOCK and T. BRIDGEMAN, University of Liverpool.
MATHEMATICAL MODELS IN SOCIAL MANAGEMENT AND LIFE SCIENCES
D. N. BURGHES and A. D. WOOD, Cranfield Institute of Technology.
MODERN INTRODUCTION TO CLASSICAL MECHANICS AND CONTROL
D. N. BURGHES, Cranfield Institute of Technology and A. DOWNS, Sheffield University.
CONTROL AND OPTIMAL CONTROL
D. N. BURGHES, Cranfield Institute of Technology and A. GRAHAM, The Open University, Milton Keynes.
TEXTBOOK OF DYNAMICS
F. CHORLTON, University of Aston, Birmingham.
VECTOR AND TENSOR METHODS
F. CHORLTON, University of Aston, Birmingham.
TECHNIQUES IN OPERATIONAL RESEARCH
VOLUME 1: QUEUEING SYSTEMS
VOLUME 2: MODELS, SEARCH, RANDOMIZATION
B. CONNOLLY, Chelsea College, University of London
MATHEMATICS FOR THE BIOSCIENCES
G. EASON, C. W. COLES, G. GETTINBY, University of Strathclyde.
HANDBOOK OF HYPERGEOMETRIC INTEGRALS: Theory, Applications, Tables, Computer Programs
H. EXTON, The Polytechnic, Preston.
MULTIPLE HYPERGEOMETRIC FUNCTIONS
H. EXTON, The Polytechnic, Preston
COMPUTATIONAL GEOMETRY FOR DESIGN AND MANUFACTURE
I. D. FAUX and M. J. PRATT, Cranfield Institute of Technology.
APPLIED LINEAR ALGEBRA
R. J. GOULT, Cranfield Institute of Technology.
MATRIX THEORY AND APPLICATIONS FOR ENGINEERS AND MATHEMATICIANS
A. GRAHAM, The Open University, Milton Keynes.
APPLIED FUNCTIONAL ANALYSIS
D. H. GRIFFEL, University of Bristol.
GENERALISED FUNCTIONS: Theory, Applications
R. F. HOSKINS, Cranfield Institute of Technology.
MECHANICS OF CONTINUOUS MEDIA
S. C. HUNTER, University of Sheffield.
GAME THEORY: Mathematical Models of Conflict
A. J. JONES, Royal Holloway College, University of London.
USING COMPUTERS
B. L. MEEK and S. FAIRTHORNE, Queen Elizabeth College, University of London.
SPECTRAL THEORY OF ORDINARY DIFFERENTIAL OPERATORS
E. MULLER-PFEIFFER, Technical High School, Ergurt.
SIMULATION CONCEPTS IN MATHEMATICAL MODELLING
F. OLIVEIRA-PINTO, Chelsea College, University of London.
ENVIRONMENTAL AERODYNAMICS
R. S. SCORER, Imperial College of Science and Technology, University of London.
APPLIED STATISTICAL TECHNIQUES
K. D. C. STOODLEY, T. LEWIS and C. L. S. STAINTON, University of Bradford.
LIQUIDS AND THEIR PROPERTIES: A Molecular and Macroscopic Treatise with Applications
H. N. V. TEMPERLEY, University College of Swansea, University of Wales and D. H. TREVENA, University of Wales, Aberystwyth.
GRAPH THEORY AND APPLICATIONS
H. N. V. TEMPERLEY, University College of Swansea.

MATHEMATICAL THEORY OF WAVE MOTION

G. R. BALDOCK, B.Sc., Ph.D.
and
T. BRIDGEMAN, B.Sc., Ph.D.
Department of Applied Mathematics and Theoretical Physics
University of Liverpool

ELLIS HORWOOD LIMITED
Publishers · Chichester
Halsted Press: a division of
JOHN WILEY & SONS
New York · Chichester · Brisbane · Toronto

First published in 1981 by

ELLIS HORWOOD LIMITED
Market Cross House, Cooper Street, Chichester, West Sussex, PO19 1EB, England

The publisher's colophon is reproduced from James Gillison's drawing of the ancient Market Cross, Chichester.

Distributors:

Australia, New Zealand, South-east Asia:
Jacaranda-Wiley Ltd., Jacaranda Press,
JOHN WILEY & SONS INC.,
G.P.O. Box 859, Brisbane, Queensland 40001, Australia

Canada:
JOHN WILEY & SONS CANADA LIMITED
22 Worcester Road, Rexdale, Ontario, Canada.

Europe, Africa:
JOHN WILEY & SONS LIMITED
Baffins Lane, Chichester, West Sussex, England.

North and South America and the rest of the world:
Halsted Press: a division of
JOHN WILEY & SONS
605 Third Avenue, New York, N.Y. 10016, U.S.A.

British Library Cataloguing in Publication Data
Baldock, G. R
Mathematical theory of wave motion. –
(Ellis Horwood series in mathematics and applications).
1. Wave-motion, Theory of
I. Title II. Bridgeman, T
531'.113'0151 QA927 80–41357
ISBN 0-85312-225-3 (Ellis Horwood Limited, Publishers)
ISBN 0-470-27113-2 (Halsted Press)

Typeset in Press Roman by Ellis Horwood Limited.
Printed in Great Britain by R. J. Acford Ltd., Chichester, West Sussex.

Table of Contents

Preface

All undergraduate students of mathematics, physics or engineering encounter the theory of wave motion at some stage in their studies. The theory may be presented as a complete set of lectures or it may occur as an integral part in a variety of courses. Many books on this subject are heavily biased towards the physics of the motion and provide little mathematical justification, whilst others consider waves in a number of physical systems as separate topics when in fact the mathematical theory is identical.

In this text we have tried to give clear mathematical definitions and to emphasise the unity of the methods and concepts applied to the wave equation and its generalisations in various physical systems. Consequently Chapter 1 considers the kinematics and dynamics of oscillators before providing the basic defining properties of waves. The wave equation is then derived for a variety of physical systems.

Chapters 2, 3 and 4 show how the one-dimensional wave equation may be solved by direct integration and also by using the various Fourier methods.

In a majority of real systems wave propagation is strongly influenced by dispersion. Thus Chapter 5 describes the fundamental characteristics and features of a dispersive medium. In addition, Chapters 6 and 10 are concerned with systems in which the dispersive effect is significant. Chapter 6 considers waves in discrete lattices and includes applications to electric filters; whilst Chapter 10 describes water waves and is used as an illustration of a physical system in which waves are propagated, but which is not characterised by the classical wave equation.

Methods of solving the wave equation in two and three dimensions are described in Chapters 7 and 8 and applied to vibrating membranes and sound waves. Similarly vector waves, which occur in electromagnetic theory, are treated in Chapter 9.

The final chapters are concerned with general hyperbolic equations, and the Cauchy problem is solved in one, two and three dimensions.

We are indebted to many of our colleagues in Liverpool who provided valuable comments on portions of the text and in particular to Dr T. A. S. Jackson and Professor G. D. Crapper for their detailed help with some of the formulations. We should also like to thank Miss Helen Wright, who did the typing of the manuscript with much patience and skill.

G. R. B., T. B.
Liverpool 1980

CHAPTER 1

Occurrence and Nature of Waves

1.1 OCCURRENCE

Any assembly of particles which mutually interact is a **medium** in which wave motion may occur. A brief disturbance in a small region induces motion in neighbouring regions, with the result that some sort of movement eventually spreads throughout the medium. This transmission of disturbance may be called **wave propagation.**

Some examples of wave motion are given in Table 1.1 In each case the interaction forces are indicated in outline. The commonly associated forms of disturbance are also shown although, in principle, any arbitrary wave form could be set in motion in any system by imposing suitable initial or source conditions. Sinusoidal motion is the simplest type of disturbance. It permeates wave theory because it is the basis for the Fourier analysis of general wave motion and also because many wave phenomena arise from periodic disturbances. We therefore begin by discussing oscillatory systems and then proceed to examine some of the features of wave motion. The chapter concludes with the derivation of the wave equation for some simple systems and consideration of some of its solutions.

1.2 OSCILLATION KINEMATICS

1.2.1 Harmonic oscillations

Any physical quantity u which depends sinusoidally on the time t,

$$u(t) = h + a\cos(\omega t + \epsilon) \quad , \tag{1.1}$$

where h, a, ω, ϵ are constants, is said to be in **harmonic oscillation**. Alternative equivalent terms are **harmonic vibration** for a coordinate of a point in a mechanical system, **monochromatic** for an electromagnetic source, and **pure tone** for an acoustic source.

The **centre of oscillation** is given by $u = h$.

The **amplitude** is a.

The **angular frequency** is ω. It is conventional to take the amplitude and angular frequency as positive quantities. For brevity ω is often called the **frequency**, but in practical applications the strict definition of frequency $f = \omega/2\pi$ is used. The **period** of the oscillation is $f^{-1} = 2\pi/\omega$.

We define the **phase** of the oscillation as $\theta(t) = -(\omega t + \epsilon)$. It can be seen from (1.1) that the sign of θ is immaterial. The minus sign is chosen to conform with the convention we shall use for travelling waves.

Table 1.1

Phenomenon	Medium	Interaction	Wave form
Ripples on a lake	Surface layer of water	Surface tension and gravity	Sinusoidal
Wind waves on water	Layer of water close to surface	Fluid pressure and gravity	Periodic
Tidal surges in estuaries	Water at all depths	Fluid pressure and gravity	Transient pulse
Transverse wave travelling on a long string	Elastic string	Tension	Arbitrary
Transverse vibrations of strings and membranes	Elastic material	Tension	Periodic
Traffic density waves	Assembly of vehicles	Mutual repulsion of drivers	Arbitrary
Seismic waves	Earth layers	Elasticity	Arbitrary
Acoustic waves	Gases, also in liquids and solids	Elasticity	Periodic
Electromagnetic waves	Vacuum or dielectric	Fundamental field	Sinusoidal
Currents in coaxial cables	Conductors and dielectric	Electromagnetic	Arbitrary

The **phase angle** is ϵ. It depends on the choice of the time origin and is of significance when the phases of two oscillations are to be compared.

The value of h depends on the origin of u. Henceforward we shall choose the origin so that $h = 0$, and so

$$u(t) = a\cos(\omega t + \epsilon) \quad . \tag{1.2}$$

For all values of a and ϵ, u satisfies the **equation of harmonic motion**

$$\ddot{u} + \omega^2 u = 0 \quad . \tag{1.3}$$

It is convenient to define the corresponding **complex oscillation**

$$\psi(t) = ae^{i\theta(t)} = ae^{-i(\omega t + \epsilon)} \quad .$$

Then ψ also satisfies (1.3). We use $\psi = Ce^{-i\omega t}$, where C is a complex constant, to stand for the real oscillation $u = \mathrm{Re}\ \psi = a\cos(\omega t + \epsilon)$, where $a = |C|$ and $\epsilon = -\arg C$.

1.2.2 Compound oscillations

A **compound harmonic oscillation** is any linear combination of harmonic oscillations of different frequencies, $\psi = \sum_{r=1}^{n} C_r e^{-i\omega_r t}$. Consider the compound oscillation

$$\psi = C_1 e^{-i\omega_1 t} + C_2 e^{-i\omega_2 t} \quad , \tag{1.4}$$

where $\omega_2 > \omega_1$. Notice that ψ does not satisfy an equation of the form (1.3), and it need not be a periodic function.

If ω_1 and ω_2 are close, that is, if

$$\omega_2 - \omega_1 \ll \omega_1 + \omega_2 \quad ,$$

it is convenient to write $\omega_1 = \omega_0 - p, \quad \omega_2 = \omega_0 + p$. Then

$$\psi = e^{-i\omega_0 t}\,(C_1 e^{ipt} + C_2 e^{-ipt})$$

$$= a(t)\, e^{i\theta(t)} \quad , \tag{1.5}$$

where

$$a(t) = |C_1 e^{ipt} + C_2 e^{-ipt}| \quad , \tag{1.6}$$

$$\theta(t) = -\omega_0 t - \epsilon(t) \quad ,$$

$$\epsilon(t) = -\arg(C_1 e^{ipt} + C_2 e^{-ipt}), \text{ and } p \ll \omega_0 \quad .$$

We can regard ψ as an oscillation with a slowly varying **instantaneous amplitude** $a(t)$. If it is an acoustic vibration **beats** will be heard, the **beat frequency** being $2p = \omega_2 - \omega_1$, as can be seen from (1.6). In the case $|C_1| = |C_2|$, $a(t)$ varies from 0 to $2|C_1|$ and the phase angle ϵ remains constant (see Fig. 1.1a); ψ is then a varying-amplitude oscillation of frequency ω_0.

Fig. 1.1b illustrates the case $|C_2| > |C_1|$. The amplitude varies between $|C_2| - |C_1|$ and $|C_2| + |C_1|$ and the phase angle also varies. The zeros of the corresponding real oscillation

$$u(t) = a(t) \cos [\omega_0 t + \epsilon(t)] \tag{1.7}$$

will be unevenly spaced. The observed frequency, estimated by counting the number of zeros of $u(t)$ in a time interval short compared with $2\pi/p$, will also vary slowly. This variable frequency is the same as the **instantaneous frequency,** which we define, for any oscillation of the form (1.5), as

$$\omega(t) = -\,\mathrm{d}\theta/\mathrm{d}t \quad .$$

For the oscillation (1.7) we therefore have $\omega(t) = \omega_0 + \mathrm{d}\epsilon/\mathrm{d}t$. Setting the wheels into motion in Fig. 1.1b we see that, when $|C_2| > |C_1|$, $\mathrm{d}\epsilon/\mathrm{d}t$ is always positive, and so ω is always greater than the mean frequency ω_0. By considering the increase in ϵ over a period $2\pi/p$ we see that the average frequency is ω_2. Similarly if $|C_2| < |C_1|$, $\omega < \omega_0$ and the average frequency is ω_1.

1.2.3 Modulation

Oscillations of varying amplitude or frequency are generated in radio transmitters in order that low-frequency (p) audio signals may be carried by high-frequency

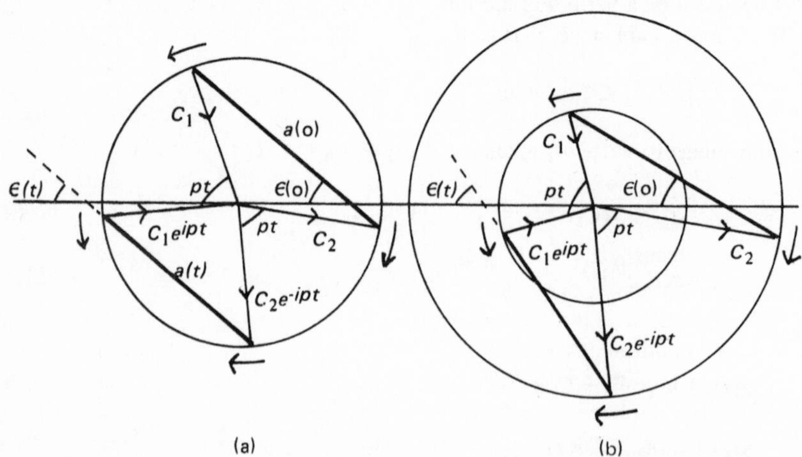

Fig. 1.1

(ω_0) radiation, the audio signals being subsequently extracted in the receiver. In the transmitter a carrier oscillation $A\cos\omega_0 t$ (A, ω_0 constant) is established and then **modulated** by the signal $S(t)$ which is slowly varying in the sense that

$$S'(t) << \omega_0 S(t) \quad .$$

In **amplitude modulation** the oscillation generated is

$$u(t) = [1 + \alpha S(t)]\ A\cos\omega_0 t \quad .$$

The maximum value of $|\alpha S(t)|$ is the **depth of modulation** d, and the constant α must be fixed so that $d < 1$ so as to prevent confusion of the signal in the receiver.

In **phase modulation** the oscillation is

$$u(t) = A\cos\,[\omega_0 t + \beta\, S(t)] \quad .$$

The instantaneous frequency is $\omega(t) = \omega_0 + \beta\, S'(t)$, and the constant β must be fixed so that $|\beta S'(t)|$ does not exceed a value depending on the bandwidth permitted by the broadcasting authority.

In **frequency modulation** the instantaneous frequency is

$$\omega(t) = \omega_0 + \gamma S(t) \quad ,$$

where the constant γ is fixed by bandwidth considerations. The oscillation is

$$u(t) = A\cos\,[\omega_0 t + \gamma \int_0^t S(\tau)\mathrm{d}\tau] \quad .$$

The amplitude is constant in both phase and frequency modulation. Both forms have the same appearance, differing only in the manner in which the signal $S(t)$ is incorporated.

The compound oscillation (1.5) is an example of an oscillation which is modulated both in amplitude and in phase (or frequency). In the case $|C_1| = |C_2|$ it is just an amplitude-modulated oscillation, and its depth of modulation is 2.

In **pulse modulation,** short bursts of oscillation at the carrier frequency are modulated in amplitude, duration, or timing by **samples** of $S(t)$ taken at regular intervals of time. Although the full signal function $S(t)$ cannot be transmitted, the method enables several streams of information to be transmitted on one carrier wave.

1.3 OSCILLATOR DYNAMICS

1.3.1 A mechanical oscillator

Let us consider a simple mechanical oscillatory system. A particle P of mass m is connected by a light spring of stiffness μ to a fixed point A on a horizontal table. The motion of the particle is resisted by a force of magnitude k times the speed, where k is a constant. An external force $F(t)$, measured in the direction AP, acts on the particle.

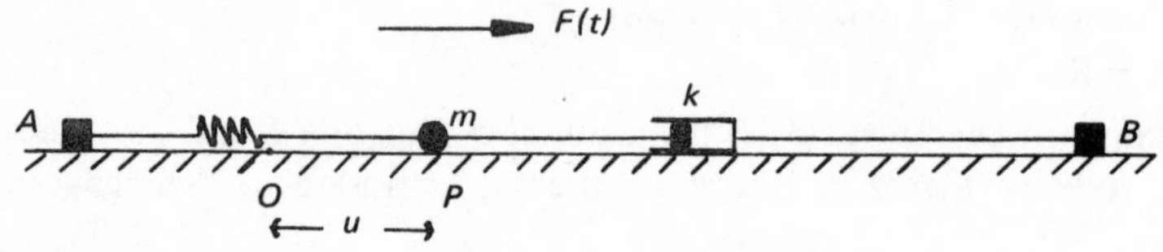

Fig. 1.2

The total resistance due to friction from all sources is symbolised in Fig. 1.2 by a single **damper** consisting of a piston moving in a fluid contained in a cylinder attached to a fixed point B. The particle can be in equilibrium at O when the spring is unstretched. At time t let $OP = u(t)$. Then the equation of motion of the particle is

$$m\ddot{u} + k\dot{u} + \mu u = F(t) \quad . \tag{1.8}$$

This is a second order inhomogeneous linear differential equation with constant coefficients. It is valid as long as $\dot{u}(t)$ and $u(t)$ remain within the **linear limits,** which are the limits of operation of the linear damping relation and Hooke's law for the spring. Many natural and artifical processes do take place within the linear limits of the relevant force systems, so (1.8) is widely applicable. It is convenient to write

$$\lambda = k/2m, \; \beta = \sqrt{(\mu/m)}, f(t) = F(t)/m \quad .$$

The equation becomes

$$\ddot{u} + 2\lambda\,\dot{u} + \beta^2 u = f(t) \quad . \tag{1.9}$$

1.3.2 Free motion

If $f(t) = 0$ the oscillator is **free**, and $u(t)$ satisfies the **homogeneous** equation

$$\ddot{u} + 2\,\lambda\dot{u} + \beta^2 u = 0 \quad . \tag{1.10}$$

This possesses the complex solution

$$\psi(t) = Ce^{-i\omega t} \quad ,$$

where C is a complex constant, provided that

$$-\omega^2 - 2i\lambda\omega + \beta^2 = 0 \quad . \tag{1.11}$$

If $\lambda < \beta$ this equation has a root of the form $\omega = \Omega - i\lambda$, where $\Omega = \sqrt{(\beta^2 - \lambda^2)} > 0$; ω is called the **complex frequency**. Then

$$\psi(t) = (A + iB)\, e^{-\lambda t}\, e^{-i\Omega t} \quad ,$$

where $A = \mathrm{Re}C$ and $B = \mathrm{Im}C$, giving

$$u(t) = e^{-\lambda t}\,(A\cos\Omega t + B\sin\Omega t)$$

or

$$u(t) = ae^{-\lambda t}\cos(\Omega t + \epsilon) \quad , \tag{1.12}$$

where a and ϵ are constants; $u(t)$ is a **damped harmonic oscillation** of frequency Ω and **decay index** $\delta = 2\pi\lambda\Omega^{-1}$. It is an amplitude-modulated oscillation; during a time interval $2\pi\Omega^{-1}$ the amplitude changes by the factor $e^{-\delta}$. The other root of (1.11), $-\Omega - i\lambda$, is not needed, because it provides no further solutions.

If $\lambda > \beta$, ω is pure imaginary and the real solutions are exponential functions. There are no oscillations and the system is said to be **overdamped.**

1.3.3 Forced motion

The **initial value problem** for the **forced motion** is:
Given the initial values $u(0)$ and $\dot{u}(0)$, find $u(t)$ to satisfy (1.9).

To solve this problem we write

$$u = v + w,$$

where $v(t)$ is a free oscillation satisfying the **given initial conditions**

$$v(0) = u(0),\ \dot{v}(0) = \dot{u}(0) \tag{1.13}$$

and $w(t)$ is that particular solution of (1.9) which satisfies **zero initial conditions**

$$w(0) = 0,\ \dot{w}(0) = 0 \quad . \tag{1.14}$$

Then $u(t)$ certainly satisfies (1.9) and the given initial conditions, as required.

From (1.12) and (1.13) we find that

$$v(t) = u(0)\, e^{-\lambda t}\cos\Omega t + [\dot{u}(0) + \lambda u(0)]\ \Omega^{-1} e^{-\lambda t}\sin\Omega t \quad . \tag{1.15}$$

We find $w(t)$ by the **method of impulses.** The idea of the method is to replace the driving function $f(t)$ by the sum of a sequence of functions:

$$f(t) = \sum_n g_n(t) \quad ,$$

where

$$\begin{aligned} g_n(t) &= f(t) \quad , \quad \tau_n < t < \tau_{n+1} \quad , \\ &= 0 \quad , \quad \text{otherwise,} \end{aligned}$$

and the support of $g_n(t)$ is a short interval I_n of length $\tau_{n+1} - \tau_n = h_n$. The graph of $g_n(t)$ is the heavy line in Fig. 1.3.

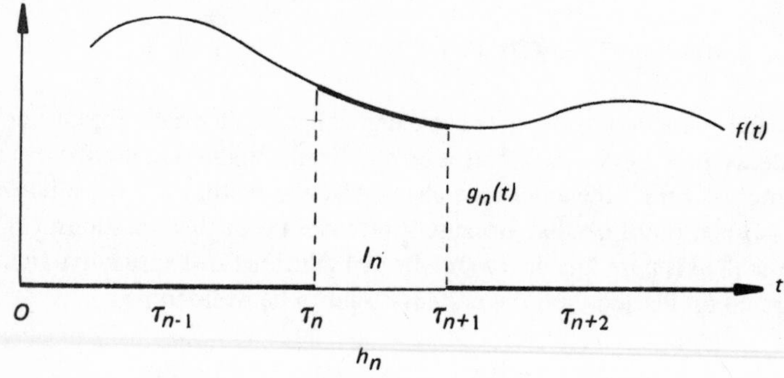

Fig. 1.3

Then $w(t) = \sum_n w_n(t)$, where $w_n(t)$ satisfies

$$\ddot{w}_n + 2\lambda\, \dot{w}_n + \beta^2 w_n = g_n(t)$$

together with $w_n(0) = 0$, $\dot{w}_n(0) = 0$.

Clearly $w_n(t)$ is zero for $t < \tau_n$; the initial conditions are therefore equivalent to $w_n(\tau_n) = 0$, $\dot{w}_n(\tau_n) = 0$.

Within the interval I_n, $g_n(t) \cong f(\tau_n)$, and so $w_n(t)$ is approximately the free oscillation produced by the application at time τ_n of an impulse $mf(\tau_n)\, h_n$ to a particle when it is at rest in equilibrium. Therefore $w(t)$ can be constructed as the sum of solutions of the homogeneous equation (1.10), the force $mf(t)$ having been replaced by a sequence of impulses applied to the free system. Proceeding to the limit of short intervals, the sums become integrals, and we arrive at the following result, a special case of **Duhamel's principle**:

Define $\phi(t,\tau)$ a function of time t and a parameter τ, as the solution of

$$\phi_{tt} + 2\lambda\, \phi_t + \beta^2 \phi = 0$$

satisfying

$$\phi(\tau,\tau) = 0,\ \phi_t(\tau,\tau) = f(\tau) \quad .$$

(Note that '$\phi d\tau$' corresponds to the solution w_n discussed above). Then the required function w is

$$w(t) = \int_0^t \phi(t,\tau)\, d\tau \quad .$$

Proof: Clearly $w(0) = 0$, as required.

Now $\quad \dot{w}(t) = \phi(t,t) + \int_0^t \phi_t(t,\tau) d\tau = 0 + \int_0^t \phi_t(t,\tau) d\tau \quad .$

Hence $\dot{w}(0) = 0$, as required,

and

$$\ddot{w}(t) = \phi_t(t,t) + \int_0^t \phi_{tt}(t,\tau) d\tau \quad ,$$

$$= f(t) + \int_0^t \phi_{tt}(t,\tau) d\tau \quad ,$$

from which it can be seen that $\ddot{w} + 2\lambda\dot{w} + \beta^2 w = f(t)$, and the proof is complete.

Using (1.15) with the origin shifted to $t = \tau$,

$$\phi(t,\tau) = \Omega^{-1} f(\tau) e^{-\lambda(t-\tau)} \sin\Omega(t-\tau) \quad .$$

Hence, writing $\tau = t - s$,

$$w(t) = \Omega^{-1} \int_0^t f(t-s) e^{-\lambda s} \sin\Omega\, s\, ds \quad . \tag{1.16}$$

1.4. FORCED OSCILLATIONS

When the driving function $f(t)$ is sinusoidal there is a sinusoidal particular solution, which is more convenient to use than (1.16). Noting that

$$\left(\frac{d^2}{dt^2} + 2\lambda\frac{d}{dt} + \beta^2\right) e^{-ipt} = (-p^2 - 2i\lambda p + \beta^2)e^{-ipt} \quad ,$$

we see that, if $f(t) = \text{Re}\, ce^{-ipt}$, (1.17)

where c is a complex constant and $p > 0$, then

$$W(t) = \text{Re}\,\frac{ce^{-ipt}}{\beta^2 - p^2 - 2i\lambda p}$$

is a particular solution. It is a harmonic oscillation of frequency p, amplitude $A = |c|r^{-1}$, where $r = \sqrt{\{(\beta^2 - p^2)^2 + 4\lambda^2 p^2\}}$ and **phase lag** $\alpha = \cos^{-1}\{(\beta^2 - p^2)r^{-1}\}$ $\alpha = \cos^{-1}\{(\beta^2 - p^2)r^{-1}\}$ relative to $f(t)$, and is called the **forced oscillation.**

The general solution of (1.9) with the driving function (1.17) consequently has the form

$$u(t) = ae^{-\lambda t}\cos(\Omega t + \epsilon) + A\cos(pt - \arg c - \alpha)$$

with a and ϵ depending on the initial conditions. This shows that, whatever the initial conditions may be, the **response** of the system to a sinusoidal driving force may be decomposed into a **transient** response (the free oscillations, which eventually decay) and the **steady state** response (the forced oscillations). If we vary the driving frequency p the amplitude A of the steady state response will behave as in Fig. 1.4a. A takes its maximum value at the **resonance frequency** $p_m = \sqrt{(\beta^2 - 2\lambda^2)}$. (For a **heavily damped** system, with $\beta^2 < 2\lambda^2$, the maximum is at zero frequency).

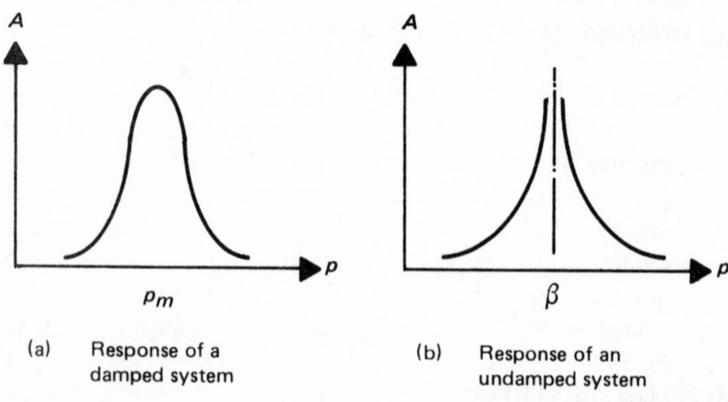

(a) Response of a damped system

(b) Response of an undamped system

Fig. 1.4

In the undamped case $\lambda = 0$, the free oscillations persist undiminished, forming with the forced oscillation a compound harmonic oscillation. As shown in Fig. 1.4b, the amplitude of the response increases indefinitely as p approaches the resonance frequency β. At resonance there is a particular solution of the form $bte^{-i\beta t}$, where b is a constant, an oscillation with a steadily growing amplitude.

1.5 AN ANALOGOUS ELECTRIC CIRCUIT

The mechanical oscillator discussed in the preceding sections is shown, slightly rearranged, in Fig. 1.5a. It has the electrical analogue shown in Fig. 1.5b, which obeys the same equation under the following correspondences.

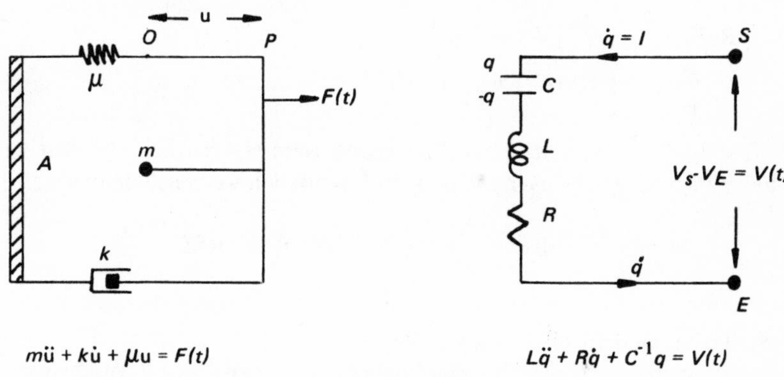

$m\ddot{u} + k\dot{u} + \mu u = F(t)$ $\qquad$ $L\ddot{q} + R\dot{q} + C^{-1}q = V(t)$

Fig. 1.5

Displacement u	Charge q
Velocity $\dot{u}$	Current $\dot{q} = I$
Driving force F	Imposed e.m.f. V
Mass m	Inductance L
Damping k	Resistance R
Spring compliance μ^{-1}	Capacitance C
Rate of working $F(t)\,\dot{u}(t)$	Rate of working $V(t)\,I(t)$

The analogy may be extended to more complicated networks of components. It has been valuable in the study of mechanical systems such as loudspeakers and shock absorbers, because the electrical analogues can easily be provided with components which can be varied so as to obtain the desired behaviour. Using either the above scheme, or a similar scheme in which velocity corresponds to potential and force to electric current, any mechanical system consisting of masses, springs and dampers can be represented by an electrical analogue. The converse, however, is not true, for instance there is no mechanical high-pass filter corresponding to the circuit shown in Chapter 6, Problem 6.6.

When the resistance alone comprises the electric circuit the quantity $V(t)/I(t)$ is equal to the constant R. In general V/I depends on t, but in the case of the forced oscillations $I = I_0 e^{-ipt}$ produced by a complex driving potential $V = V_0 e^{-ipt}$, where I_0 and V_0 are complex constants, it can be seen from the differential equation that $V = IZ(p)$, where the function $Z(p)$, called the **complex impedance** of the circuit, is defined by $Z(p) = V_0/I_0$. In this case $Z(p) = -iLp + R + iC^{-1}p^{-1}$.

Likewise we define the **mechanical impedance** of the system in Fig. 1.5a in terms of a driving force $F = F_0 e^{-ipt}$ producing forced oscillations $u = u_0 e^{-ipt}$. $F = \dot{u}Z(p)$, where $Z(p) = -imp + k + i\mu p^{-1}$. We shall see that impedance is a useful concept in the discussion of the interactions at the boundary of a medium in which waves are being transmitted.

The instantaneous rate of working of the real driving force ReF is

$$\begin{aligned} S(t) = \operatorname{Re}F \operatorname{Re}\dot{u} &= \tfrac{1}{4}\,(F + \bar{F})\,(\dot{u} + \bar{\dot{u}}) \\ &= \tfrac{1}{2}\operatorname{Re}(F\bar{\dot{u}}) + \tfrac{1}{2}\operatorname{Re}(F\dot{u}) \end{aligned}$$

The first term is constant, while the second term is a harmonic oscillation of frequency $2p$. Averaging over one period of oscillation we conclude that

$$\text{Mean power dissipation} = \tfrac{1}{2}\operatorname{Re}(F\bar{\dot{u}}) = \tfrac{1}{2}\,|\dot{u}|^2 \operatorname{Re}Z \quad .$$

1.6 NORMAL MODES OF VIBRATION

Consider a system described by coordinates $q_1(t), q_2(t), \ldots, q_n(t)$ and governed by linear equations of motion of the form

$$\sum_{j=1}^{n} (m_{ij}\ddot{q}_j + \mu_{ij}q_j) = 0, i = 1, 2, \ldots, n, \tag{1.18}$$

where m_{ij} and μ_{ij} are constants. A convenient way of solving these equations is to begin by seeking solutions of the form

$$q_j(t) = Q_j \cos(\omega t + \epsilon) \quad , \tag{1.19}$$

where ω, ϵ and Q_j are constants. Evidently the constants Q_j must satisfy the linear homogeneous algebraic equations

$$\sum_{j=1}^{n} (\mu_{ij} - \omega^2 m_{ij})\,Q_j = 0, i = 1, 2, \ldots, n. \tag{1.20}$$

The determinant of these equations, which is a polynomial of degree n in ω^2, vanishes for at most n positive values of ω. These are called the **normal frequencies** of the system. To each of them will correspond at least one set of ratios $Q_1 : Q_2 : \ldots : Q_n$, which are called the **normal modes** of the system, and a solution such as (1.19) is called a **normal vibration.** It can be shown (Goult 1978) that, if the mass matrix (m_{ij}) and the stiffness matrix (μ_{ij}) are both symmetric and positive definite (always so in systems oscillating about equilibrium at all $q_j = 0$), then the normal frequencies are real and there are exactly n linearly independent normal modes. Further, the normal vibrations form a **basis** for all motions of the system, that is, any motion can be expressed in the form

$$q_j(t) = \sum_{k=1}^{n} \alpha_k Q_j^{(k)} \cos(\omega_k t + \epsilon_k), j = 1, 2, \ldots, n \quad , \qquad (1.21)$$

in which α_k and ϵ_k are constants to be determined from initial conditions, $Q_1^{(k)} : Q_2^{(k)} : \ldots : Q_n^{(k)}$ is the kth normal mode and ω_k) is the corresponding normal frequency.

In effect the system has been replaced by an equivalent system of independent simple oscillators.

This idea is extensively used later in connection with continuous systems.

1.7 PROPAGATING WAVES

Consider a scalar physical quantity u which depends on a single space coordinate x and on the time t. If, in some domain of xt-space, u can be expressed in the form

$$u(x, t) = f(x - ct), \text{ where } c \text{ is constant,} \qquad (1.22)$$

then u is said to be a **wave** which **propagates** in the x-direction with velocity c. In Fig. 1.6, the diagrams (a) and (b) are pictures of the wave taken at $t = 0$ and at a later instant $t = t_1$.

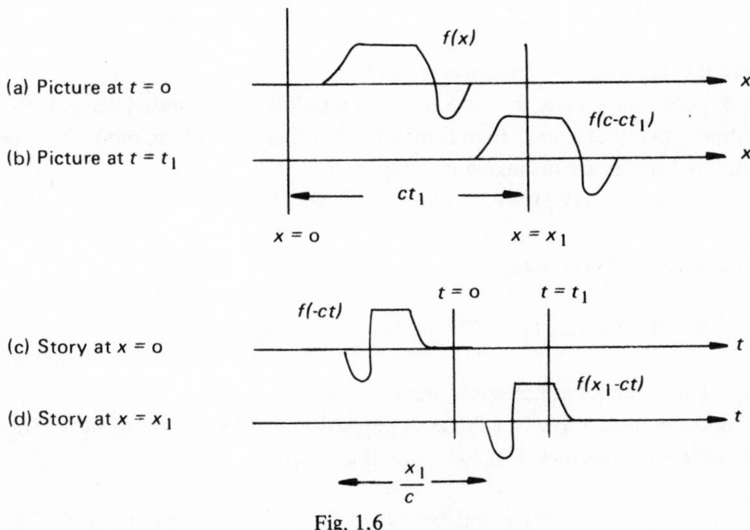

Fig. 1.6

The wave moves along with constant velocity c (positive in this illustration) without changing its shape. If c were negative, the wave would propagate to the left. Diagrams (c) and (d) show that the behaviour of u at $x = x_1 > 0$ is an exact replica of the behaviour at $x = 0$, but delayed by the time interval $c^{-1}x_1$.

From (1.22) we see that, for any shape f,

$$c(\partial u/\partial x) = -(\partial u/\partial t) \quad . \tag{1.23}$$

This is therefore the partial differential equation of waves with velocity c. Waves of the form $u = g(x + ct)$ travel with velocity $-c$ and satisfy the equation $c\partial u/\partial x = \partial u/\partial t$. Every wave travelling with speed c in either direction therefore satisfies the one-dimensional **classical wave equation** $c^2 \dfrac{\partial^2 u}{\partial x^2} = \dfrac{\partial^2 u}{\partial t^2}$. (1.24)

All the definitions in section 1.2 on oscillations can be extended to waves. A **harmonic wave** is a function

$$u(x, t) = a \cos(kx - \omega t - \epsilon), \; (a > 0, \omega > 0),$$

in which the constant k is called the **wave number** or **propagation number.** The **wavelength** is $2\pi k^{-1}$. The velocity $c = \omega/k$, and the wave travels in the positive x direction if $k > 0$.

The **phase** of the wave is $\theta(x, t) = kx - \omega t - \epsilon$.

We may define the corresponding **complex harmonic wave**

$$\psi(x, t) = ae^{i\theta(x,t)} = Ce^{i(kx - \omega t)}$$

where $u = \text{Re}\,\psi$, $a = |C|$ and $\epsilon = -\arg C$.

If a depends on x or t, u is said to be an **amplitude-modulated wave.** If $\theta(x, t)$ is non-linear in either x or t, then u may be described as a **phase-modulated wave.** Modulated waves are discussed in Chapter 5.

The function $u(x,t) = e^{-\lambda t} \cos(kx - \omega t - \epsilon), \lambda > 0,$ (1.25)

is a **damped harmonic wave.**

The function $u(x, t) = e^{-px} \cos(kx - \omega t - \epsilon), \quad p > 0,$ (1.26)

is called an **attenuated harmonic wave.**

A **compound wave** is a linear superposition of two or more waves travelling with different velocities. For example, the function

$$2\cos(x - 2t) + \cos(3x - t)$$

is a compound harmonic wave. It is not itself a propagating wave, and has no single velocity associated with it. Since the wave equation (1.24) is linear, the compound wave

$$u(x, t) = f(x - ct) + g(x + ct) \quad , \tag{1.27}$$

where f and g are arbitrary functions, is a solution; it is shown in Chapter 2 that every solution can be expressed in this form.

1.8 STATIONARY WAVES

A function of the form

$$u(x, t) = X(x)\, T(t)$$

is called a **stationary wave.** As an example, the behaviour of the stationary wave

$$u(x, t) = \sin\pi x \cos\pi t \tag{1.28}$$

is illustrated in Fig. 1.7.

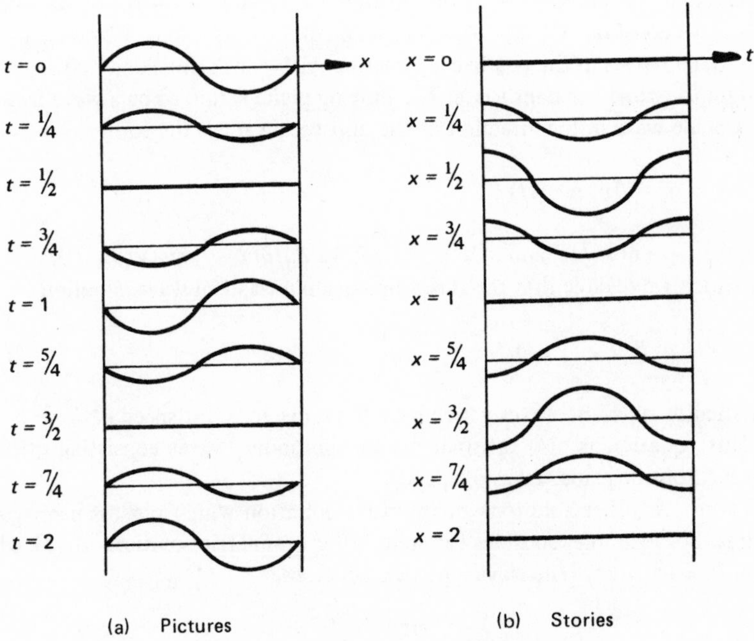

Fig. 1.7

Diagram (a) shows that the general features of the waves are constant in time; the **nodes** ($u = 0$) N_1, N_2, N_3 etc. and the **antinodes** ($u_x = 0$) A_1, A_2, etc. retain their position permanently. Diagram (b) shows that all points oscillate in the same, or exactly opposite, phase; the amplitude of the motion at the point x being $|\sin \pi x|$. Such motion is thus better described as an oscillation or vibration.

The function u in this case is a solution of the equation $u_{xx} = u_{tt}$, and therefore must be expressible as a compound wave. Indeed,

$$u = \sin\pi x \cos\pi t = \tfrac{1}{2} \sin\pi(x - t) + \tfrac{1}{2} \sin\pi(x + t) \quad , \tag{1.29}$$

which is a superposition of waves travelling in opposite directions with unit speed.

Linear combinations of stationary waves are used to construct solutions of linear partial differential equations to satisfy given initial and boundary conditions. This method, called the Fourier method, is developed in Chapter 3 and is used in many applications.

1.9 WAVES IN THREE DIMENSIONS

A function $u(\mathbf{r},t)$ defined on a region of three dimensional space during some time interval is a **plane wave** in the positive x-direction if $u = f(x - ct)$, where c is a positive constant. On any plane perpendicular to the x-axis, u is constant, and if the plane moves in the positive x-direction with constant velocity c then u will maintain the same constant value. The moving plane is said to be a **wave front.**

A plane wave in the direction of the unit vector $\mathbf{n}$ has the form

$$u = f(\mathbf{r} \,.\, \mathbf{n} - ct) \quad .$$

Then $c\nabla u = -\mathbf{n}(\partial u/\partial t)$ and $c^2\nabla^2 u = -c\nabla.(\mathbf{n}\ \partial u/\partial t) = -c\mathbf{n}.\ \nabla(\partial u/\partial t)$, from which we deduce that the **three-dimensional classical wave equation**

$$c^2\nabla^2 u = \partial^2 u/\partial t^2 \tag{1.30}$$

is satisfied by all plane waves which have the same constant speed c.

This equation is also satisfied by all compound waves consisting of linear superpositions of plane waves of speed c.

There are other solutions of the wave equation which possess propagating features. For example, consider the spherically symmetric solutions $u(r, t)$, where $r = \sqrt{(x^2 + y^2 + z^2)}$. The wave equation becomes

$$\frac{1}{r^2}\frac{\partial}{\partial r}\left(r^2 \frac{\partial u}{\partial r}\right) = \frac{1}{c^2}\frac{\partial^2 u}{\partial t^2} \quad .$$

Defining $v(r, t) = ru(r, t)$ we find that the equation simplifies to

$$c^2 v_{rr} = v_{tt}.$$

Hence there are solutions of the form

$$u = \frac{f(r - ct)}{r} \quad ,$$

where f is any function. The behaviour of u on the sphere $r = r_2$ is a replica of its behaviour on the sphere $r = r_1$, where $r_1 < r_2$, reduced by the factor r_2/r_1 and delayed by the time interval $c^{-1}(r_2 - r_1)$. That is, a recognisable disturbance is propagated outwards, so we can call u a **spherical wave.** The spheres on which $u = 0$ expand with constant speed c, and so may be termed wave fronts. Other features, such as the points at which u takes a nonzero constant value, do not in general move with constant speed.

1.10 WAVES IN STRETCHED STRINGS

The equilibrium state of an elastic string that is stretched in a straight line by longitudinal forces may be specified by the **density** $\rho(x)$ (mass per unit length), **elastic modulus** $\lambda(x)$, and **tension** $T_0(x)$. The tension depends on the forces which are maintaining the equilibrium, but we are not concerned with these forces. We shall study both the free motion of the string and the forced motion produced by additional external forces.

1.10.1 Longitudinal motion

We shall assume that the string is subject to **Hooke's Law** throughout the motion. If a segment of the string with natural length h, sufficiently short for $\lambda(x)$ to be considered constant, is stretched to a length h_1, then the tension T in the segment

is given by $T_1 = \lambda \dfrac{h_1 - h}{h}$,

whence $h(T_1 + \lambda) = \lambda h_1$.
If the same segment is at tension T_2 when its length is h_2, then

$$h(T_2 + \lambda) = \lambda h_2 \quad .$$

Hence

$$\frac{T_2 + \lambda}{T_1 + \lambda} = \frac{h_2}{h_1} \quad . \tag{1.31}$$

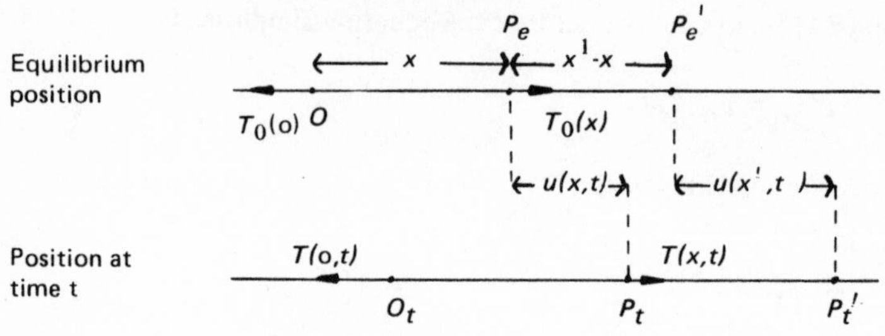

Fig. 1.8

Fig. 1.8 shows the string first in its equilibrium position and then in its displaced position at time t. Let O be a fixed origin and let P_e, P_e' be the equilibrium positions and P_t, P_t' be the positions at time t of neighbouring points P, P' of the string.

Let $OP_e = x$, $OP_e' = x'$ and $OP_t = x + u(x, t)$, so that $OP_t' = x' + u(x', t)$. Hence $P_e P_e' = x' - x$ and $P_t P_t' = x' - x + u(x', t) - u(x, t)$.

Then, from (1.31)

$$\frac{T(x, t) + \lambda(x)}{T_0(x) + \lambda(x)} \cong \frac{P_tP_t'}{P_eP_e'} = 1 + \frac{u(x', t) - u(x, t)}{x' - x} .$$

Taking limits as $P_e P_e' \to 0$, we obtain

$$\frac{T(x, t) + \lambda(x)}{T_0(x) + \lambda(x)} = 1 + \frac{\partial u}{\partial x} ,$$

which may be written

$$T(x, t) - T_0(x) = [T_0(x) + \lambda(x)] \; \partial u/\partial x \quad . \tag{1.32}$$

Suppose that the external forces consist of those needed to maintain the equilibrium distribution of tension $T_0(x)$ together with an additional force density $g(x, t)$. By this we mean that the short segment $P_e P_e'$ is subject to an extra force approximately equal to $g(x, t)\, P_e P_e'$. Then the equation of motion of the portion $O_t P_t$ of the string is obtained by equating the sum of all the extra forces to the total rate of change of momentum.

$$T(x, t) - T(0, t) - [T_0(x) - T_0(0)] + \int_0^x g(\zeta, t)\mathrm{d}\zeta = \int_0^x \rho(\zeta)u_{tt}(\zeta, t)\mathrm{d}\zeta \quad .$$

Substituting for $T(x, t)$ from (1.32) and differentiating with respect to x we obtain

$$\{[T_0(x) + \lambda(x)]\,u_x\}_x + g(x,\, t) = \rho(x)u_{tt} \quad . \tag{1.33}$$

In the case when $g(x, t) = 0$, this equation becomes the **equation of free motion** $[(T_0 + \lambda)u_x]_x = \rho(x)u_{tt}$. When $g = 0$ and $\rho(x)$, $T_0(x)$ and $\lambda(x)$ are all constant, the equation reduces to the form of the classical wave equation $c^2u_{xx} = u_{tt}$. Therefore a uniform evenly-stretched string transmits longitudinal waves with speed

$$c = \sqrt{\left(\frac{T_0 + \lambda}{\rho}\right)} \quad . \tag{1.34}$$

Equation (1.33) is valid provided that the string does not become slack. The solution $u(x, t)$ must therefore be restricted so that (1.32) gives values of the tension $T(x, t)$ which are positive at all values of x and t within the domain under consideration.

The extra force density function may not be known explicitly. For example, $g(x, t)$ might be the sum of an explicitly given **driving term** $f(x, t)$ and a **linear damping term** $-Ru_t$, where R is constant.

1.10.2 Other elastic systems

The **longitudinal motion of a helical spring** obeys the same equation. In this case $T(x, t)$ may take negative values. A uniform unstretched spring transmits longitudinal waves with speed $\sqrt{(\lambda/\rho)}$.

Similarly a **uniform unstressed thin rod** of volume density ρ_V and Young's modulus E transmits longitudinal waves with speed $\sqrt{(E/\rho_V)}$.

The **torsional motion of a thin rod** provides another example of a system governed by an equation of the form of (1.33). The equation in this case is

$$[C(x)\,u_x]_x + G(x,\, t) = J(x)u_{tt} \quad ,$$

where $u(x, t)$ is the **angular displacement** from equilibrium of a cross-section at a point P,

$C(x)$ is the **torsional rigidity** of the rod, which is the moment of the couple which would maintain in equilibrium unit angular displacement per unit length,

$G(x, t)$ and $J(x)$ are respectively the moment of the external forces and the moment of inertia, per unit length, about the line of centroids of the cross-sections.

When $G = 0$ and C and J are constant the rod transmits waves with a speed depending on the shape of the cross-section; when the cross-section is circular the wave speed is $\surd(\mu/\rho_V)$, where μ is the rigidity modulus (Landau and Lifshitz 1959).

1.10.3 Small transverse motion of a stretched string.

We shall assume that the string is perfectly flexible, supporting no shear stresses, and that the internal force system consists solely of the tension T. In Fig. 1.9 the x-axis is the configuration which the string can take when it is in equilibrium at tension $T_0(x)$ under some longitudinal force system, P_e being the equilibrium position of the point P of the string.

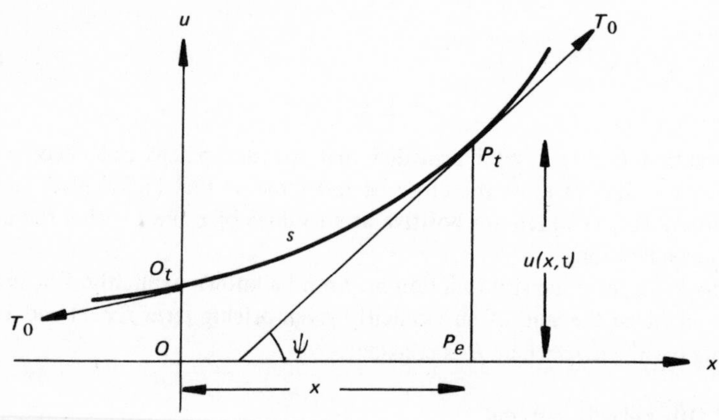

Fig. 1.9

We consider **small** transverse motion confined to a plane. At time t the point P of the string is at P_t which has coordinates $[x, u(x, t)]$ referred to axes through O. We shall say that the motion is small when $(\partial u/\partial x) \ll 1$ always, that is, the angle ψ which the tangent to the string makes with the x-axis is always small. This restriction carries several implications.

(i) The x-component of the tension $T\cos\psi \cong T$,

(ii) the transverse component $T\sin\psi \cong T\tan\psi = Tu_x$,

(iii) the fractional arc increment $(\partial s/\partial x) - 1 = \sec\psi - 1 \cong \frac{1}{2}\psi^2$, which is small.

Adapting (1.32) to the present notation, we see that

$$\frac{T}{T_0} - 1 = \left(1 + \frac{\lambda}{T_0}\right)\left(\frac{\partial s}{\partial x} - 1\right) .$$

The right-hand side is small provided that $T_0/\lambda \gg \psi^2$, which does not severely restrict T_0 because ψ itself is small. We may therefore take $T(x, t) \cong T_0(x)$ at all times, and the postulated transverse motion can take place without the imposition of additional external longitudinal forces.

We may now write down the equation of small transverse motion of the segment O_tP_t when it is subject to a **transverse force density** $g(x, t)$:

$$T_0(x)u_x(x, t) - T_0(0)u_x(0, t) + \int_0^x g(\zeta, t)\mathrm{d}\zeta = \int_0^x \rho(\zeta)u_{tt}(\zeta, t)\mathrm{d}\zeta .$$

Differentiating with respect to x we obtain

$$\{T_0(x)u_x\}_x + g(x, t) = \rho(x)u_{tt} \quad . \tag{1.35}$$

The **free transverse motion of a uniformly stretched uniform string** is therefore subject to the equation

$$c^2u_{xx} = u_{tt}, \text{ where } c = \sqrt{(T_0/\rho)} \quad . \tag{1.36}$$

We have seen that the speed of longitudinal waves in the same string is $\sqrt{\left(\frac{T_0 + \lambda}{\rho}\right)}$. In a slightly stretched string $T_0 \ll \lambda$ and the longitudinal waves are much faster than the transverse waves. This situation occurs in a violin string. In a weak string T_0 may exceed λ and the speeds of longitudinal and transverse waves are of the same order; a weak coiled spring is an example.

1.10.4 Vector waves on a string

The above treatment is easily extended to general small transverse displacements $\mathbf{u}(x, t)$, where $\mathbf{u}$ is a two dimensional vector in the plane perpendicular to the x-axis. The result is equation (1.35) with u and g replaced by two-dimensional vectors. Equation (1.36) is then replaced by a **vector wave equation**

$$c^2\mathbf{u}_{xx} = \mathbf{u}_{tt} \quad . \tag{1.37}$$

A **plane polarized wave** is a wave of form $\mathbf{u}(x, t) = \mathbf{n}u(x, t)$ where $\mathbf{n}$ is a fixed unit vector. The **direction of polarization** is $\mathbf{n}$ and the **plane of polarization** is the plane containing $\mathbf{n}$ and the direction of propagation (along the x-axis in this case). Every wave satisfying (1.37) may be expressed as a linear superposition of two waves with mutually perpendicular directions of polarization.

Harmonic waves with velocity c in the positive direction have the form

$$\mathbf{u} = \mathbf{A}\cos(kx - kct) + \mathbf{B}\sin(kx - kct) \quad ,$$

where $\mathbf{A}$ and $\mathbf{B}$ are constant vectors. At a fixed point x, the extremity of the vector $\mathbf{u}$ describes generally an ellipse during each period $2\pi/(kc)$; the wave is therefore said to be **elliptically polarized**. In the special case when $\mathbf{A}$ and $\mathbf{B}$ are parallel the wave is plane polarized. When $|\mathbf{A}| = |\mathbf{B}|$ and $\mathbf{A}.\mathbf{B} = 0$ the wave is **circularly polarized.** By suitably agitating one end of a long rope it is easy to generate waves with any type of polarization.

1.11 THE LOSSLESS TRANSMISSION LINE

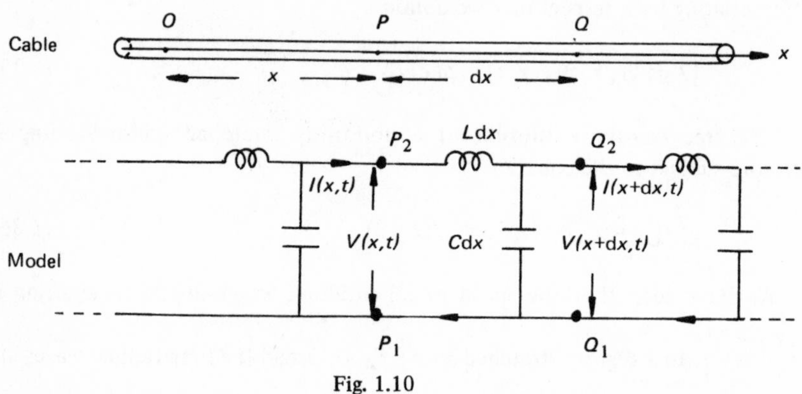

Fig. 1.10

A telegraph cable may be treated as the limit of a sequence of elementary circuits composed of inductances, capacitances and resistances. In the ideal lossless line there are no resistances. In Fig. 1.10 the section of the cable between $P(x)$ and $Q(x + \mathrm{d}x)$ is represented by an elementary circuit consisting of a series inductance $L\mathrm{d}x$ and a shunt capacitance $C\mathrm{d}x$, where L and C are constants. At P the current passing the terminals P_1, P_2 is $I(x, t)$ and the potential difference between the terminals is $V(x, t)$. The circuit equations are

$$\frac{\partial I}{\partial x} + C\frac{\partial V}{\partial t} = 0 \ , \qquad \frac{\partial V}{\partial x} + L\frac{\partial I}{\partial t} = 0 \ , \tag{1.38}$$

from which it follows that $V_{xx} = LC\, V_{tt}$. The cable therefore transmits voltage fluctuations at the speed $c = (LC)^{-1/2}$.

Harmonic waves of frequency $\omega > 0$ can propagate. They may be represented by

$$V = V_0 e^{i(kx - \omega t)} \ , \quad I = I_0 e^{i(kx - \omega t)} \ . \tag{1.39}$$

Let us momentarily consider the general case of a line which contains resistances. It is found in Chapter 5 that k is then complex, so that the waves are attenuated and proceed with velocity $\omega/\mathrm{Re}k$. We define the **characteristic impedance** of the line at frequency ω as

$$Z(\omega) = V_0/I_0 \ ,$$

when a wave of the form (1.39), where Rek and ω are positive, is being transmitted. In the sense in which impedance is defined in section 1.5, $Z(\omega)$ is the impedance of a semi-infinite line considered as a circuit with input terminals P_1, P_2, as shown in Fig. 1.11a. The relation between V and I at the input is thus unaffected, at frequency ω, if the whole line is replaced by a single circuit of impedance $Z(\omega)$ connected to P_1 and P_2. Also, if the line is terminated at a point R by connecting $Z(\omega)$ to the output terminals R_1 and R_2, then waves of frequency ω will proceed along the line as if the line were semi-infinite. We can thus construct a line which is effectively infinite at a given frequency.

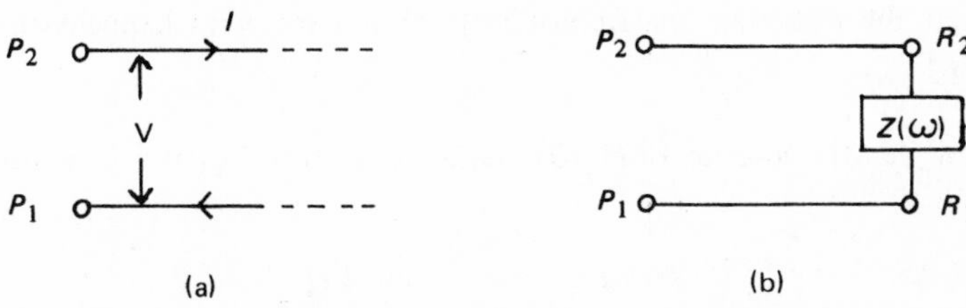

Fig. 1.11

Returning now to the lossless line we deduce from (1.38) that $Z(\omega)$ is constant and equal to $\sqrt{(L/C)}$, the positive root being taken because of the assumption $k > 0$. The semi-infinite lossless line thus behaves like a pure resistance. In this case the ratio V/I is constant also for arbitrary waves, provided we agree to the condition that $V = 0$ when $I = 0$, as can be seen by substituting $I = f(x - ct)$ in (1.38).

The power delivered at the input is $VI = I^2Z$. This is transmitted along the line and is equal to the **energy flux,** which is the rate of flow of energy past a fixed point. If the line is terminated by a resistance Z this energy is entirely dissipated in the resistance; we say that the terminating impedance is **matched** to the line.

If the line is terminated at R by a circuit with impedance not equal to Z the appropriate solutions of (1.38) are compound waves, some of the energy flux being reflected back along the line. The impedance presented at the terminals P_1 and P_2 then becomes dependent on ω and on the length PR (see problem 1.11).

PROBLEMS

1.1 A point P moves so that its acceleration is directed towards a fixed point O and is of magnitude $\omega^2 OP$. Writing $\vec{OP} = \mathbf{u}$, show that

$$\mathbf{u} = \mathbf{A}\cos\omega t + \mathbf{B}\sin\omega t,$$

where $\mathbf{A}$ and $\mathbf{B}$ are constants. Express $\mathbf{u}$ in the form $\mathbf{u} = \text{Re}\,\boldsymbol{\psi}$, where $\boldsymbol{\psi} = \mathbf{C}e^{-i\omega t}$ and $\mathbf{C}$ is a complex constant, and show that

(i) the motion takes place in a plane perpendicular to $i\mathbf{C} \times \bar{\mathbf{C}}$,
(ii) the motion is rectilinear if $\mathbf{C} \times \bar{\mathbf{C}} = 0$,
(iii) the motion is circular if $\mathbf{C} \cdot \mathbf{C} = 0$,
(iv) generally P moves in an ellipse with principal axes in the directions $\mathrm{Re}\, \mathbf{C}e^{-i\alpha}$, $\mathrm{Im}\, \mathbf{C}e^{-i\alpha}$, where $\alpha = \frac{1}{2} \arg (\mathbf{C} \cdot \mathbf{C})$,
(v) OP^2 varies between the limits $\frac{1}{2}(\mathbf{C} \cdot \bar{\mathbf{C}} \pm |\mathbf{C} \cdot \mathbf{C}|)$.

1.2 The function $u = \cos \Omega t$, where $\Omega = \omega_0 (1 + \alpha \cos pt)$ and α and $p\omega_0^{-1}$ are small, looks as if it might describe a frequency-modulated wave. Explain why it is unsatisfactory.
[Work out the instantaneous frequency $\omega(t)$ and see what happens to it as t increases.]

1.3 Show that the solution of (1.10), subject to $u(0) = a$, $\dot{u}(0) = b$, is, for $\lambda < \beta$,

$$u(t) = e^{-\lambda t} \left[a \cos \Omega t + \frac{\lambda a + b}{\Omega} \sin \Omega\, t\right] .$$

1.4 A measure Q of the sharpness of the resonant response of the system in section 1.4 is obtained by considering the frequencies p_1, p_2 at which the amplitude of the forced oscillation is $1/\sqrt{2}$ of the amplitude at the resonant frequency p_m. We define $Q = \left|\dfrac{2p_m}{p_1 - p_2}\right|$. Show that, for a lightly damped system ($\lambda \ll \beta$), $Q \cong \beta/\lambda$.

1.5 Consider the equation $\ddot{u} + 2\lambda\, \dot{u} + u = 0$ subject to $u(0) = 1$ and $\dot{u}(0) = 0$: a damped oscillator initially at rest at unit deflection from equilibrium. Show that, for $\lambda < 1$, $|u(t)| = e^{-\lambda t}$ whenever $|u(t)|$ has a maximum. Compare the values of $e^{-\lambda t}$, when $\lambda = \frac{1}{2}$ and $t = 5$, 10 and 20, with the quantities $u(5)$, $u(10)$ and $u(20)$ in the cases $\lambda = 1$ **(critical damping)** and $\lambda = \frac{3}{2}$.
[This example illustrates the fact that if an instrument contains parts which are capable of vibrating it settles down most quickly after a disturbance if it is critically damped.]

1.6 **Coupled oscillators.** The equations

$$\ddot{u}_1 + \beta^2 u_1 = \gamma^2 (u_2 - u_1),\; \ddot{u}_2 + \beta^2 u_2 = \gamma^2 (u_1 - u_2)$$

represent two identical interacting oscillators. Show that the normal frequencies of vibration are given by $\omega_1^2 = \beta^2$, $\omega_2^2 = \beta^2 + 2\gamma^2$ and that the amplitudes of the corresponding normal modes are in the ratios 1:1 and 1:–1.

Show that the initial conditions

$$u_1(0) = 0, \dot{u}_1(0) = 2, u_2(0) = 0, \dot{u}_2(0) = 0$$

lead to the solution

$$\begin{pmatrix} u_1 \\ u_2 \end{pmatrix} = \frac{\sin \omega_1 t}{\omega_1} \pm \frac{\sin \omega_2 t}{\omega_2} \quad .$$

In the case $\gamma << \beta$ (**weak coupling**) write this as

$$\beta^{-1} (\sin \omega_1 t \pm \sin \omega_2 t) + (\text{a small term})$$

and show that in this motion $u_1(t)$ and $u_2(t)$ exhibit beats which are approximately in complementary phase (that is, the energy passes slowly to and fro between the two oscillators).

1.7 Solve the equation

$$\ddot{w} + 4\dot{w} + 13w = e^{-2it}, w(0) = 0, \dot{w}(0) = 0$$

by using Duhamel's principle and (1.16).

1.8 In the case of undamped resonance,

$$\ddot{u} + \beta^2 u = ce^{-i\beta t} \quad ,$$

find the constant b in the particular solution $bte^{-i\beta t}$ and describe the phase of the forced oscillation relative to the driving force.

1.9 A wave $g(x + ct)$ $(c > 0)$ is given by

$$\begin{aligned} g(\zeta) &= \sin \zeta, 0 < \zeta < \pi/2 \quad , \\ &= 0 \qquad , \text{otherwise} \end{aligned}$$

Draw the shape of the wave at the instants $t = 0$, $t = \pi/c$ and also a graph of the motion at the points $x = 0, x = -\pi$.

1.10 Verify that $r^{-1}f(r - ct)$ and $r^{-1}g(r + ct)$ are solutions of the wave equation (1.30). Choosing $f(\zeta)$, $g(\zeta)$ as $e^{\pm ik\zeta}$, obtain stationary spherical waves by linear superposition.

1.11 The transmission line of section 1.11 is terminated by joining R_1 to R_2. Express V as a compound wave $Ae^{i(kx-\omega t)} + Be^{-i(kx+\omega t)}$ and find I, and hence the impedance $Z(\omega)$, at P when $PR = h$.

CHAPTER 2

Solutions of the Wave Equation by Integration

2.1 THE GENERAL SOLUTION

We now derive some general results which may be applied to all systems characterised by the one-dimensional wave equation

$$c^2 \frac{\partial^2 u}{\partial x^2} = \frac{\partial^2 u}{\partial t^2} \quad . \tag{2.1}$$

We transform to new variables ξ and η defined by

$$\xi = x - ct, \; \eta = x + ct.$$

Then

$$2x = \eta + \xi, \;\; 2ct = \eta - \xi.$$

Writing $u(x, t) = v(\xi, \eta)$, we have $2cv_\xi = cu_x - u_t$, and hence

$$4c^2 v_{\xi\eta} = c^2 u_{xx} + cu_{xt} - cu_{tx} - u_{tt} = c^2 u_{xx} - u_{tt} \quad .$$

The wave equation thus becomes $v_{\xi\eta} = 0$,

which has the general solution $v = f(\xi) + g(\eta)$, giving

$$u = f(x - ct) + g(x + ct) \quad , \tag{2.2}$$

where f and g are arbitrary C^2 functions.

We have thus shown that every solution of the one-dimensional classical wave equation is a superposition of two waves of unchanging shape travelling in opposite directions, each with constant speed c.

2.1.1 D'Alembert's solution of the initial value problem

When the general solution is used to solve a particular problem the functions f and g are determined from the initial conditions and any boundary conditions which are imposed on the system. We now consider the problem:

Find $u(x, t)$ satisfying $c^2 u_{xx} = u_{tt}$, $|x| < \infty$, $t > 0$ subject to the conditions

$$u(x, 0) = a(x) \text{ and } u_t(x, 0) = b(x) \quad , \tag{2.3}$$

where $a(x)$ and $b(x)$ are given for $|x| < \infty$.

Substituting from (2.3) into (2.2), we obtain the relations

$$a(x) = f(x) + g(x) \quad , \tag{2.4}$$

$$b(x) = cf'(x) - cg'(x) \quad . \tag{2.5}$$

Note that since f and g are to be C^2 functions a solution of the problem exists only if

$$a \text{ is a } C^2 \text{ function and } b \text{ is a } C^1 \text{ function} \quad . \tag{2.6}$$

Integrating (2.5), we have

$$\frac{1}{c} \int_{x_0}^{x} b(s) \, ds = f(x) - g(x) \quad , \tag{2.7}$$

where x_0 is an arbitrary constant. From (2.4) and (2.7) we determine $f(x)$ and $g(x)$:

$$\begin{aligned} f(x) &= \tfrac{1}{2}a(x) - \frac{1}{2c} \int_{x_0}^{x} b(s)ds \\ g(x) &= \tfrac{1}{2}a(x) + \frac{1}{2c} \int_{x_0}^{x} b(s)ds \quad . \end{aligned} \tag{2.8}$$

Substituting these expressions into the general solution (2.2), we obtain

$$u(x, t) = \tfrac{1}{2}[a(x - ct) + a(x + ct)] + \frac{1}{2c} \int_{x-ct}^{x+ct} b(s)ds \quad . \tag{2.9}$$

Equation (2.9) is known as **d'Alembert's solution** of the one-dimensional wave equation. The interval $[x_1 - ct_1, x_1 + ct_1]$ $(t_1 > 0)$ of the x-axis is called the **interval of dependence**, or **domain of dependence** of the point (x_1, t_1), because $u(x_1, t_1)$ depends only on the values taken by a at the ends of the interval and the values taken by b at all points in the interval.

The region $\mathfrak{D}$ of the xt-plane in which $t > 0, x - ct \leqslant x_1$ and $x + ct \geqslant x_1$ is called the **region of influence** of the point x_1, because the value of a at x_1 influences the solution $u(x, t)$ on the boundary of $\mathfrak{D}$, and the value of b at x_1 influences $u(x, t)$ throughout $\mathfrak{D}$.

2.1.2 'Infinite strings'

In this chapter solutions of the wave equation in various cases will be illustrated by means of the transverse motion of a string (analysed in section 1.10.3). Although the d'Alembert solution (2.9) has been derived in the domain $|x| < \infty$, it is applicable in a limited sense to a real string occupying the interval $L_1 < x < L_2$. Suppose the initial functions $a(x)$ and $b(x)$ are zero everywhere except within an interval $X_1 < x < X_2$. Then (2.9) is valid in the interval $0 \leqslant t < T$, where cT is the lesser of $X_1 - L_1$ and $L_2 - X_2$. During this time interval the ends of the string are irrelevant, and it is customary to describe the system as an 'infinite string'.

2.1.3 Discontinuities in the initial functions

In physical problems, $a(x)$ and $b(x)$ usually satisfy the differentiability conditions (2.6), but for the sake of simplicity we often approximate by using functions which do not satisfy them, for example the function $a(x)$ shown in Fig. 2.1c. The d'Alembert result is then used as a formula to produce a corresponding 'solution'. This is an acceptable device, for it is easily shown that the solution of the initial-value problem is a continuous function of the initial conditions. Specifically, for any $\epsilon > 0$ and any time t in an arbitrary finite interval $[t_0, t_1]$, we can find an $\eta > 0$ such that if we replace $a(x)$ and $b(x)$ by $\bar{a}(x)$ and $\bar{b}(x)$ where

$$|a(x) - \bar{a}(x)| < \eta \quad , \quad |b(x) - \bar{b}(x)| < \eta \quad ,$$

and if $\bar{u}(x, t)$ is the d'Alembert 'solution' corresponding to $\bar{a}(x)$, $\bar{b}(x)$, then $|u(x, t) - \bar{u}(x, t)| < \epsilon$ (see Problem 2.1). Hence by a suitable choice of $a(x)$, $b(x)$ the difference between $u(x, t)$ and $\bar{u}(x, t)$ can be made arbitrarily small.

2.1.4 Initial velocity everywhere zero

In this case $b(x) = 0$ and from (2.9) we obtain

$$u(x, t) = \tfrac{1}{2}\,[a(x - ct) + a(x + ct)] \quad ,$$

from which it is easy to compute the value of u at any point (x, t). Suppose, for example, that $a(x)$ has the shape shown in Fig. 2.1c, taking non-zero values only in an interval (α, β). The forward and backward waves are both initially equal to $\tfrac{1}{2}a(x)$, (Fig. 2.1a, b), and to find the solution at time t we displace these graphs a distance ct in opposite directions and then sum the ordinates of the displaced graphs. Fig. 2.1(d to h) illustrates the solution at different values of t.

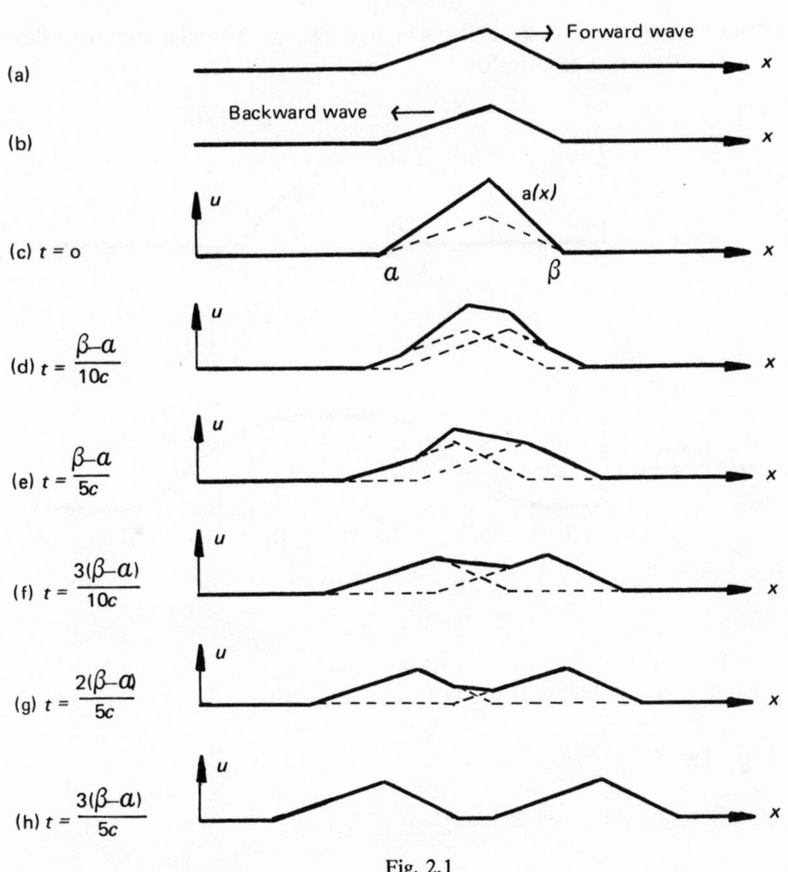

Fig. 2.1

2.1.5 Initial displacement everywhere zero

In this case $a(x) = 0$ and the solution is given by

$$u(x, t) = \frac{1}{2c}\int_{x-ct}^{x+ct} b(s)\mathrm{d}s \quad .$$

As an example, suppose $b(x)$ has the shape shown in Fig. 2.2a, taking non-zero values only in the interval (α, β), and let us seek the solution at some time $t > (\beta - \alpha)/2c$. We consider five regions:

$$\text{I: } x \leqslant \alpha - ct, \quad \text{II: } \alpha - ct \leqslant x \leqslant \beta - ct,$$
$$\text{III: } \beta - ct \leqslant x \leqslant \alpha + ct, \quad \text{IV: } \alpha + ct \leqslant x \leqslant \beta + ct, \quad \text{V: } \beta + ct \leqslant x \quad .$$

When x lies in region I, $x + ct < \alpha$. The entire interval of integration then lies outside the interval (α, β), and hence $u(x, t) = 0$. A similar argument shows that $u(x, t) = 0$ when x is in region V.

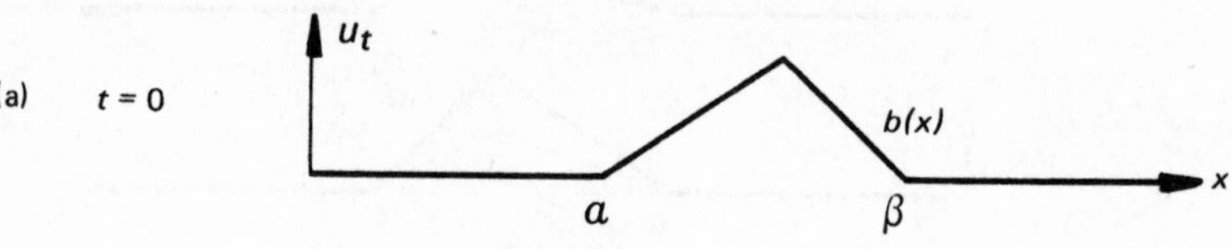

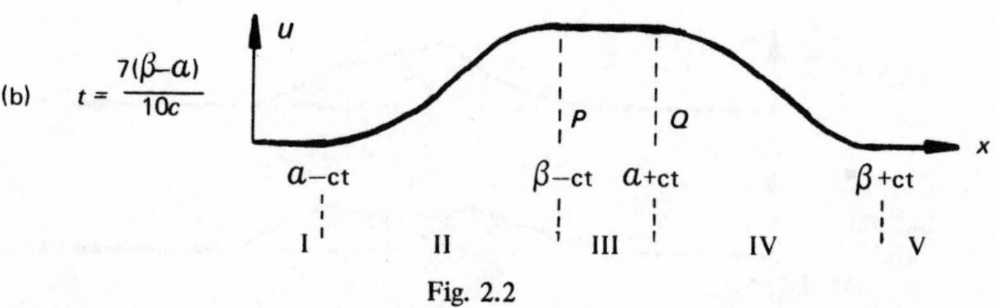

Fig. 2.2

For values of x in region II we have

$$\alpha < x + ct < \beta \quad .$$

Also, since $t > (\beta - \alpha)/2c$ we deduce that

$$x - ct < \alpha \quad .$$

Therefore

$$u(x, t) = \frac{1}{2c}\int_{\alpha}^{x+ct} b(s)ds \quad .$$

Similarly, when x lies in region IV, the solution is

$$u(x, t) = \frac{1}{2c}\int_{x-ct}^{\beta} b(s)ds \quad .$$

Finally, when x lies within the third region,

$$x - ct < \alpha \text{ and } x + ct > \beta;$$

so the interval (α, β) is entirely contained within the interval $(x - ct, x + ct)$. Consequently

$$u(x, t) = \frac{1}{2c}\int_{\alpha}^{\beta} b(s)ds \quad ;$$

that is, $u(x, t)$ is constant. Thus, when $t > (\beta - \alpha)/2c$, the solution may be represented by the curve in Fig. 2.2b.

When $t < (\beta - \alpha)/2c$ the result is modified because the points P and Q in Fig. 2.2b appear in the reverse order, and so the expression for $u(x, t)$ in the region between P and Q is

$$u(x, t) = \frac{1}{2c}\int_{x-ct}^{x+ct} b(s)ds \quad ,$$

which is not constant.

2.2 WAVE ENERGY

Since the general solution shows that waves in both directions are propagated with constant speed and unvarying shape, it is natural to expect that the wave energy, once it is given to the system, is propagated without loss.

To illustrate the propagation of energy in a one-dimensional system we consider the particular case of small transverse vibrations of a uniform string of density ρ and constant tension T. The kinetic energy possessed by that part of the string from $x = x_0$ to $x = x_1$ is equal to

$$\tfrac{1}{2}\rho\int_{x_0}^{x_1} u_t^2\, dx \quad ,$$

whilst the potential energy is the work done by the tension in the stretching consequent on the displacement from the equilibrium configuration. The length of an elementary section δx of the string becomes $\sqrt{(1 + u_x^2)}\delta x$ in a displaced configuration and so the potential energy is

$$T\int_{x_0}^{x_1} [\sqrt{(1 + u_x^2)} - 1]\, dx \quad .$$

This, to second order, is

$$\tfrac{1}{2}T\int_{x_0}^{x_1} u_x^2\, dx \quad ,$$

and so the total energy of the segment $[x_0, x_1]$ of the string is $W(x_0, x_1)$, where

$$W(x_0, x_1) = \tfrac{1}{2}\rho \int_{x_0}^{x_1} (u_t^2 + c^2 u_x^2)\, \mathrm{d}x \quad .$$

Consequently, if we define $\mathcal{E}(x, t)$ to be energy density at a point x at time t in such a way that $\mathcal{E}(x, t)\,\delta x$ is the total energy in the infinitesimal interval $[x, x+\delta x]$, then

$$\mathcal{E}(x, t) = \tfrac{1}{2}\rho\, u_t^{\,2} + \tfrac{1}{2}\, T u_x^{\,2} \quad .$$

The rate of change of the energy of the segment is

$$\begin{aligned}
\frac{\mathrm{d}W(x_0, x_1)}{\mathrm{d}t} &= \rho \int_{x_0}^{x_1} (u_t u_{tt} + c^2 u_x u_{xt})\, \mathrm{d}x \\
&= T \int_{x_0}^{x_1} (u_t u_{xx} + u_x u_{xt})\, \mathrm{d}x \\
&= T \int_{x_0}^{x_1} \frac{\partial}{\partial x} (u_t u_x)\, \mathrm{d}x \quad . \\
&= T u_t(x_1, t)\, u_x(x_1, t) + [-T u_t(x_0, t)\, u_x(x_0, t)] \quad .
\end{aligned}$$

The first term on the right-hand side represents the energy flow into the segment from the region $x > x_1$ and the second term represents the flow into the segment from the region $x < x_0$.
Hence

$$\frac{\mathrm{d}W(x_0, x_1)}{\mathrm{d}t} = \mathcal{F}(x_0, t) - \mathcal{F}(x_1, t),$$

where
$$\mathcal{F}(x, t) = -T u_x u_t \tag{2.10}$$

is the instantaneous value of the energy flux passing through the point $P(x)$ at time t. This can be seen in another way by observing that $-Tu_x$ is the transverse force which the part of the string to the left of P exerts on the part to the right of P, and u_t is the transverse velocity of P (see Fig. 2.3). Hence $-Tu_x u_t$ is the rate at which the left-hand part works on the right-hand part, that is, the rate of transfer of energy. In this respect the terms $-Tu_x$ and u_t are analogous to the voltage and current at a point in a transmission line, the product of the voltage and current being equal to the power transmitted (see section 1.11).

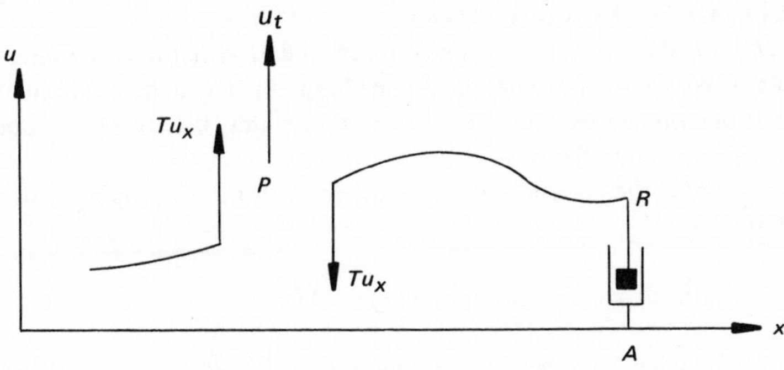

Fig. 2.3

2.3 CHARACTERISTIC IMPEDANCE

The analogy between a string and a transmission line may be continued, to define the mechanical impedance of the string which corresponds to the electrical impedance of the line. We make the following correspondences, similar to those established in section 1.5.

Voltage V	Transverse driving force from the left (see Fig. 2.3)	$-Tu_x$
Current I	Transverse velocity	u_t .

Like the lossless transmission line of section 1.11, the string transmits waves of the form $u(x, t) = f(x - ct)$, where f is an arbitrary function. Then the ratio

$$-\frac{Tu_x}{u_t} = \frac{-Tf'}{-cf'} = \frac{T}{c} = \sqrt{(T\rho)} = Z \text{ (constant)} \quad . \tag{2.11}$$

The constant Z is called the **characteristic impedance** of the string. Z is the mechanical impedance (defined in section 1.5), presented to a driving force acting at a point P, of the semi-infinite string to the right of P. It is equal to the impedance of a damper of strength Z, which exerts a resisting force of magnitude Z times the speed, connecting P to a fixed point.

Also, if the string is terminated at a point R which can move transversely subject to a damper of strength Z connected to a fixed point A (see Fig. 2.3), then waves will proceed along the string as if the string were semi-infinite.

Other types of boundary conditions at R give compound wave solutions; the impedance at P is then a function $Z(\omega)$ depending on the frequency and also on the length PR.

2.4 REFLECTION AT A BOUNDARY

Since any medium ultimately comes to an end it is natural to enquire what happens when a wave travelling in the medium reaches such an end point. The answer is determined by how the medium terminates, that is, by the boundary conditions imposed at the end.

We consider the equation $c^2u_{xx} = u_{tt}$ $(x < 0,\, t > 0)$ subject to the initial conditions

$$u(x, 0) = a(x),\ u_t(x, 0) = b(x)\ (x < 0) \quad . \tag{2.12}$$

Before introducing boundary conditions, let us denote by $v(x, t)$ the solution of the problem in the case when the functions $a(x)$ and $b(x)$ are extended to the entire real domain by defining

$$a(x) = b(x) = 0\ (x > 0) \quad .$$

Using (2.8), this solution may be written

$$v(x, t) = f(x - ct) + F(x + ct) \quad ,$$

where

$$f(\zeta) = \tfrac{1}{2}a(\zeta) + \frac{1}{2c}\int_{\zeta}^{0} b(s)\mathrm{d}s$$

and

$$F(\zeta) = \tfrac{1}{2}a(\zeta) - \frac{1}{2c}\int_{\zeta}^{0} b(s)\mathrm{d}s \quad .$$

The functions $f(\zeta)$ and $F(\zeta)$ are uniquely determined by $a(x)$ and $b(x)$, and vice versa. We note that $f(\zeta)$ and $F(\zeta)$ are both zero for $\zeta > 0$.

Suppose that, when boundary conditions are imposed at $x = 0$, the solution is

$$u(x, t) = v(x, t) + G(x - ct) + g(x + ct)\ (x < 0, t > 0) \quad ,$$

where the functions $G(\zeta)$ and $g(\zeta)$ are to be found. In order to preserve the initial conditions (2.12), both $G'(\zeta)$ and $g'(\zeta)$ must be zero when $\zeta < 0$, and consequently $G(x - ct)$ is constant for $x < 0$ and $t > 0$. We may therefore discard G and write

$$u(x, t) = f(x - ct) + F(x + ct) + g(x + ct)\ (x < 0, t > 0) \quad ,$$

where $g(\zeta) = 0$ for $\zeta < 0$ because of the initial conditions. The function $g(\zeta)$ is determined from the boundary condition, which will be a condition on the function $u(0, t)$, that is, on the function $f(-ct) + g(ct)$, because $F(ct) = 0$ for

$t > 0$. The wave $F(x + ct)$, which is thus unconnected with the boundary conditions, is the **receding wave.** We shall henceforward omit it from all calculations, and take

$$u = f(x - ct) + g(x + ct) \; (x < 0, t > 0) \tag{2.13}$$

with

$$f(\zeta) = 0 \; (\zeta > 0) \text{ and } g(\zeta) = 0 \; (\zeta < 0) \quad . \tag{2.14}$$

We call $f(x - ct)$ the **incident wave** and $g(x + ct)$ the **reflected wave.**

We now consider the transverse motion of a string occupying the region $x \leqslant 0$, subject to the four types of boundary condition shown in Fig. 2.4. In each case we find the reflected wave corresponding to a given incident wave.

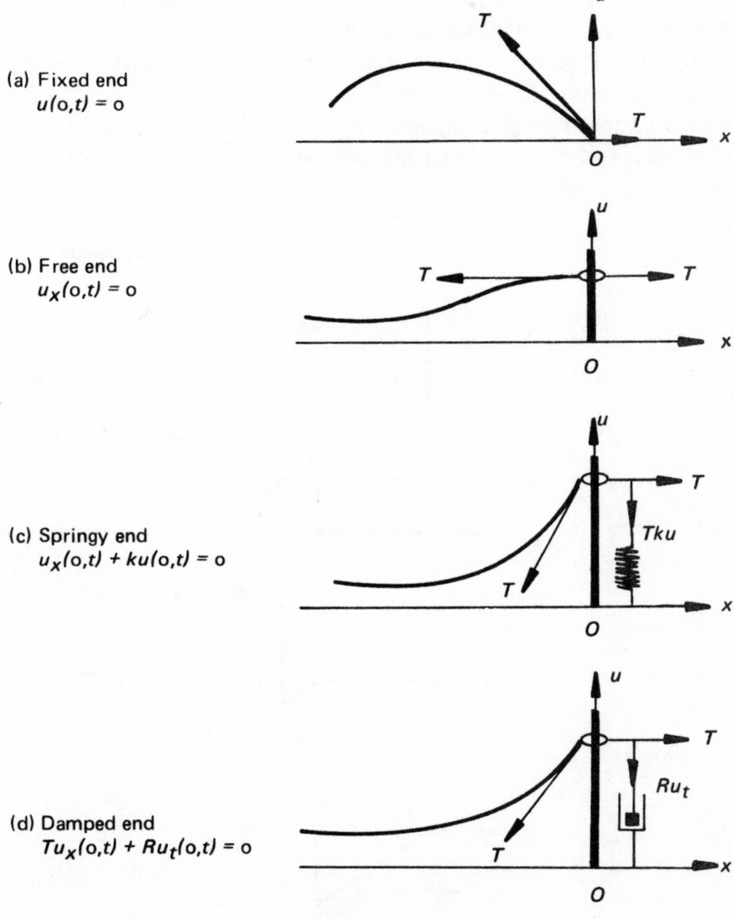

Fig. 2.4

2.4.1 Fixed end

Let the string be fastened to a fixed support at $x = 0$ (Fig. 2.4a) so that the boundary condition is $u(0, t) = 0$. Using the incident and reflected wave combination (2.13) we obtain $f(-ct) + g(ct) = 0$ ($t > 0$), giving the functional relation $g(\zeta) = -f(-\zeta)$ ($\zeta > 0$).

Hence the reflected wave is

$$g(x + ct) = -f(-x - ct) \quad ,$$

and the required solution is

$$u(x, t) = f(x - ct) - f(-x - ct) \quad . \tag{2.15}$$

In Fig. 2.5 the thick solid line to the left of O represents the string. The broken lines represent the incident and reflected waves which are added together to obtain the displacement. To simplify the diagrams an example has been chosen in which $f(\zeta) = 0$ for $\zeta > -ct_0$. Then, for $t < t_0$, the solution consists only of the incident wave $f(x - ct)$, which travels towards O. The reflected wave $-f(-x - ct)$ appears at time t_0 and travels away from O. In our example $f(\zeta)$ is also zero for $\zeta < -ct_1$, so that the incident wave disappears at time t_1, and only the reflected wave remains.

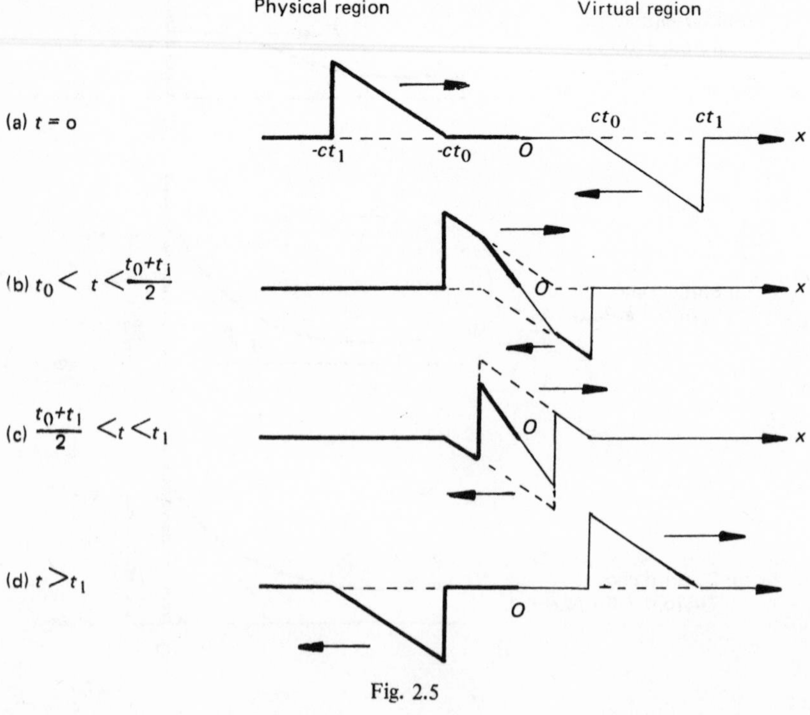

Fig. 2.5

The visualisation of the reflection process can be assisted by supposing the string to be extended into the **virtual region** $x > 0$. We may then think of the reflected wave as a wave approaching O from the virtual region. The incident and reflected waves, which have the same shape but opposite sign, move towards each other and coalesce in such a way that $u(0, t)$ is always zero. The reflected wave then continues travelling to the left in the physical string, and the incident wave travels away from O in the virtual string. In Fig. 2.5 the virtual string is indicated by the thin solid line.

2.4.2 Free end

If the string terminates in a small loop which is free to slide along a smooth rod fixed transversely at $x = 0$ (Fig. 2.4b), the boundary condition is $u_x(0, t) = 0$, and the end is said to be free.

Then $f'(-\zeta) + g'(\zeta) = 0$,

whence $g(\zeta) = f(-\zeta) + 0$,

the constant of integration being zero to satisfy (2.14).

Hence $\quad u(x, t) = f(x - ct) + f(-x - ct)$

and the incident wave is reflected back without change of form and without change of sign, as shown in Fig. 2.6.

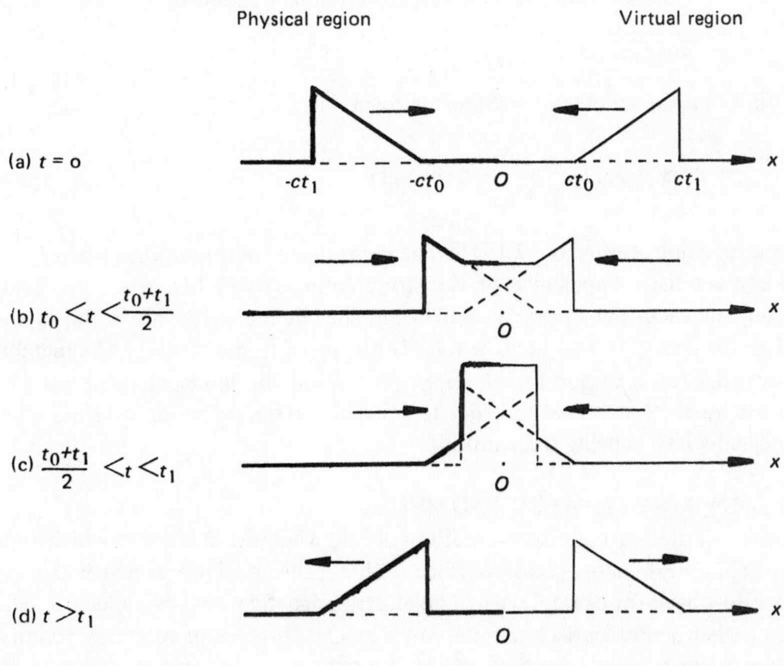

Fig. 2.6

2.4.3 Springy end

If the loop at $x = 0$ is attached to a spring which provides a transverse restoring force equal to Tk times the displacement from the equilibrium position (Fig. 2.4c), the boundary condition is $Tku(0, t) + Tu_x(0, t) = 0$.

Inserting the solution (2.13) we obtain

$$kf(-ct) + kg(ct) + f'(-ct) + g'(ct) = 0 \quad ,$$

whence

$$\frac{d}{d\zeta}\left\{ e^{k\zeta} g(\zeta) \right\} = e^{2k\zeta} \frac{d}{d\zeta}\left\{ e^{-k\zeta} f(-\zeta) \right\} \quad .$$

Integration by parts and substitution into (2.13) gives

$$u(x, t) = f(x - ct) + f(-x - ct) - 2ke^{-k(x+ct)} \int_0^{x+ct} e^{ks} f(-s)\mathrm{d}s \quad , \tag{2.16}$$

in which the lower limit of the integral has been chosen so as to satisfy (2.14).

2.4.4 Damped end

If the loop is attached to a damper which provides a transverse resisting force equal to R times the speed (Fig. 2.4d), the boundary condition is

$$Ru_t(0, t) + Tu_x(0, t) = 0 \quad .$$

We find that the reflected wave has the form

$$g(x + ct) = \frac{T - Rc}{T + Rc} f(-x - ct) \quad . \tag{2.17}$$

The interesting feature of (2.17) is that the reflected wave vanishes when $R = Z$, the characteristic impedance of the string defined by (2.11). When this occurs the impedance of the string is said to be matched by the impedance of the damper, and all the energy of the incident wave is dissipated in the damper. The matching of impedances is important in telegraphy where the impedances of electrical circuits must be matched in order to prevent a reflected pulse interfering with the signal which is being transmitted.

2.5 BOUNDARY BETWEEN TWO MEDIA

A different kind of boundary condition occurs when the medium in which a wave is travelling terminates at an interface with another medium in which the wave may also travel; for example, when an electromagnetic wave in air meets a dielectric or when sound waves in air meet an obstacle. This type of boundary condition is essentially different because whilst it produces a reflected wave in the first medium it also gives rise to a transmitted wave in the second medium.

2.5.1 Junction of two strings

Consider two semi-infinite strings S_1 and S_2, of linear densities ρ_1, ρ_2, joined at $x = 0$ and stretched at tension T, S_1 occupying the region $x < 0$ and S_2 the region $x > 0$. As the two strings have different linear densities it follows that they also have different wave speeds c_1 and c_2.

Let $f(x - c_1t)$ be a given incident wave in S_1, and let S_2 be initially undisturbed, so that $u(x, 0) = 0$ and $u_t(x, 0) = 0$ for $x > 0$. Then the wave in S_2, which may be written $h(x - c_2t) + H(x + c_2t)$, must satisfy $h'(\zeta) = 0$ and $H'(\zeta) = 0$ for $\zeta > 0$, and hence $H(x + c_2t)$ is a constant for $t > 0$. Therefore we may discard H and write

$$u(x, t) = \begin{Bmatrix} f(x - c_1t) + g(x + c_1t)\ , & x < 0 \\ h(x - c_2t) & , x > 0 \end{Bmatrix} . \tag{2.18}$$

We call $h(x - c_2t)$ **the transmitted wave.**

We note that

$$f(\zeta) = h(\zeta) = 0\ (\zeta > 0) \text{ and } g(\zeta) = 0\ (\zeta < 0) \quad . \tag{2.19}$$

There are two boundary conditions at the junction $x = 0$. The first is the geometrical condition that the displacement must be continuous:

$$u(-0, t) = u(+0, t), \tag{2.20}$$

where $u(-0, t)$ denotes $\lim_{x\to 0} u(x, t)\ (x < 0)$

and $u(+0, t)$ denotes $\lim_{x\to 0} u(x, t)\ (x > 0)$.

The second is the dynamical condition that the transverse force must be continuous:

$$u_x(-0, t) = u_x(+0, t) \quad ; \tag{2.21}$$

this condition is necessary because a non-zero resultant force acting on the infinitesimally small mass at O would produce an infinite acceleration.

Inserting the solution (2.18) into (2.20) and (2.21) gives

$$f(-c_1t) + g(c_1t) = h(-c_2t) \tag{2.22}$$

and

$$f'(-c_1t) + g'(c_1t) = h'(-c_2t) \quad .$$

Integrating we obtain

$$-\frac{1}{c_1}\ f(-c_1t) + \frac{1}{c_1}\ g(c_1t) = -\frac{1}{c_2} h(-c_2t) + 0 \quad , \tag{2.23}$$

the constant of integration being zero to satisfy (2.19).

Then, from (2.22) and (2.23),

$$g(\zeta) = \frac{c_2 - c_1}{c_2 + c_1} f(-\zeta) \tag{2.24}$$

and

$$h(\zeta) = \frac{2c_2}{c_2 + c_1} f\left(\frac{c_1}{c_2}\zeta\right) . \tag{2.25}$$

Hence the reflected and transmitted waves are given by

$$g(x + c_1 t) = \frac{c_2 - c_1}{c_2 + c_1} f(-x - c_1 t) \quad ,$$

$$h(x - c_2 t) = \frac{2c_2}{c_2 + c_1} f\left(\frac{c_1}{c_2} x - c_1 t\right) .$$

Note that, when $c_2 = 0$,

$$g(x + c_1 t) = -f(-x - c_1 t)$$

and $\quad h(x - c_2 t) = 0 \quad .$

So, by comparison with (2.15), this is equivalent to $x = 0$ being a fixed end to the string S_1. Similarly when c_2 is large the system is almost equivalent to $x = 0$ being a free end to the string S_1. In this case it is interesting to note that the amplitude of the transmitted wave cannot exceed twice that of the incident wave.

2.5.2 Energy transfer

An important feature of wave propagation is the transfer of energy through the supporting medium. We now consider what happens to the energy when a wave is reflected and transmitted at the junction. From (2.10) the energy flux in S_1 is

$$\mathcal{F} = - T u_x u_t = - T(f' + g')(-c_1 f' + c_1 g') \qquad (x < 0)$$

$$= T c_1 f'^2 (x - c_1 t) - T c_1 g'^2 (x + c_1 t)$$

$$= \mathcal{F}_{\mathrm{I}} - \mathcal{F}_{\mathrm{R}} \quad ,$$

where $\mathcal{F}_{\mathrm{I}}$ and $\mathcal{F}_{\mathrm{R}}$ may be termed the **incident flux** and the **reflected flux** respectively. The **transmitted flux** is $\mathcal{F}_{\mathrm{T}} = T c_2 h'^2 (x - c_2 t)$.

From (2.24) and (2.25) we see that the junction has **reflectivity**

$$\frac{\mathscr{F}_R}{\mathscr{F}_I} = \left(\frac{c_2 - c_1}{c_2 + c_1}\right)^2 \quad \text{and } \textbf{transmissivity} \quad \frac{\mathscr{F}_T}{\mathscr{F}_I} = \frac{4c_1c_2}{(c_2 + c_1)^2} \quad .$$

We note that $\mathscr{F}_R + \mathscr{F}_T = \mathscr{F}_I$ in accordance with energy conservation.

2.5.3 Impedance matching

In practical systems it is often important that the reflected wave should be suppressed, so that $\mathscr{F}_R = 0$. For example, if a join in a submarine cable or a telephone line produces a reflected wave it will interfere with the signal. The means of suppressing the reflected wave depends to a great extent on the physical system under consideration, and sometimes the suppression is only effective for waves of a particular frequency (see Problem 9.3). However in the case of two long strings joined at $x = 0$ it is possible to suppress the reflected wave for all forms of the incident wave if $c_1 < c_2$. This is done by inserting a damping device of strength R at the junction $x = 0$. From Fig. 2.7a we see that the dynamical boundary condition (2.21) is then replaced by

$$Tu_x(+0, t) - Tu_x(-0, t) = Ru_t(0, t) \quad .$$

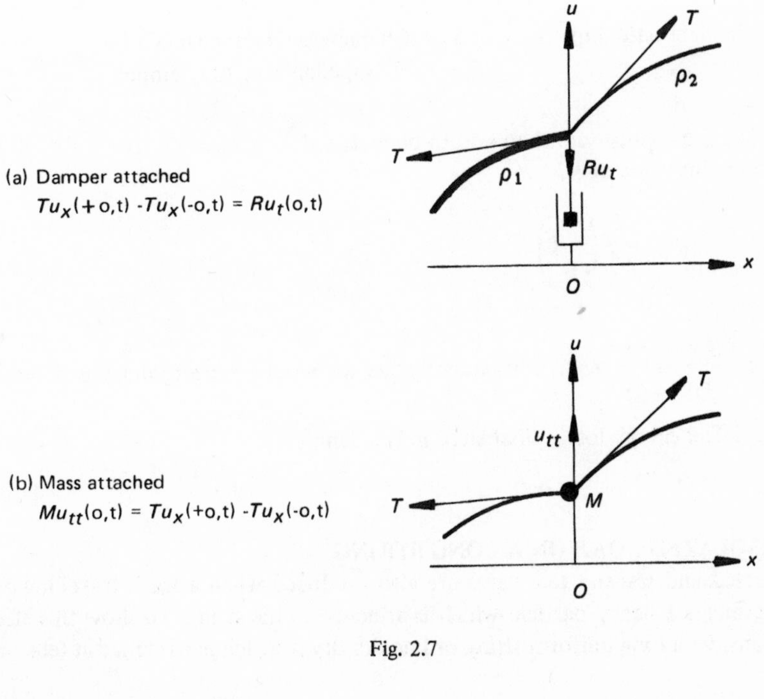

(a) Damper attached
$Tu_x(+0,t) - Tu_x(-0,t) = Ru_t(0,t)$

(b) Mass attached
$Mu_{tt}(0,t) = Tu_x(+0,t) - Tu_x(-0,t)$

Fig. 2.7

Hence

$$f'(-c_1t) + g'(c_1t) = h'(-c_2t) + c_2 \frac{R}{T} h'(-c_2t) \quad .$$

Integrating and using (2.19) gives

$$-\frac{1}{c_1} f(-c_1t) + \frac{1}{c_1} g(c_1t) = -\left(\frac{1}{c_2} + \frac{R}{T}\right) h(-c_2t) \quad . \tag{2.26}$$

Then from (2.22) and (2.26),

$$\left(\frac{1}{c_1} + \frac{1}{c_2} + \frac{R}{T}\right) g(\zeta) = \left(\frac{1}{c_1} - \frac{1}{c_2} - \frac{R}{T}\right) f(-\zeta) \quad ,$$

and so the reflected wave is suppressed when

$$R = \frac{T}{c_1} - \frac{T}{c_2} = \sqrt{T}\,(\sqrt{\rho_1} - \sqrt{\rho_2}) = Z_1 - Z_2 \quad ,$$

following the notation of (2.11). This result may be written

Characteristic impedance of S_1 = Characteristic impedance of S_2 + Impedance of the damper.

The impedances are then said to be matched.

For this value of R,

$$h(\zeta) = f\left(\frac{c_1}{c_2}\zeta\right)$$

and so $\dfrac{\mathcal{F}_\mathrm{T}}{\mathcal{F}_\mathrm{I}} = \dfrac{c_1}{c_2} < 1$. Consequently, as we would expect, there is a loss of energy. The energy lost is dissipated in the damper.

2.6 ISOLATED LOAD ON A LONG STRING

Reflected and transmitted waves are also produced when a wave travelling on a string meets a heavy particle which is attached to the string. To show this effect we consider a long uniform string of line density ρ which is stretched at tension T

and carries a particle of mass M at $x = 0$. Then, following (2.18), we suppose

$$u(x, t) = \begin{Bmatrix} f(x - ct) + g(x + ct), & x < 0 \\ h(x - ct) \qquad\qquad , & x > 0 \end{Bmatrix} .$$

The continuity of the string again gives

$$u(-0, t) = u(+0, t) \quad ,$$

and the second boundary condition is provided by the equation of motion of the mass, namely

$$Mu_{tt}(0, t) = Tu_x(+0, t) - Tu_x(-0, t)$$

(see Fig. 2.7b).

Inserting the expressions for $u(x, t)$ into the boundary conditions gives

$$f(\zeta) + g(-\zeta) = h(\zeta) \tag{2.27}$$

$$\text{and } c^2M\, h''(\zeta) = T[h'(\zeta) - f'(\zeta) - g'(-\zeta)] .$$

Integrating and choosing the constant to satisfy (2.19) we obtain

$$Mh'(\zeta) = \rho\,[h(\zeta) - f(\zeta) + g(-\zeta)] + 0 \quad .$$

Eliminating g, $\qquad Mh'(\zeta) = 2\rho\ [h(\zeta) - f(\zeta)]$.

The required solution of this differential equation is

$$h(\zeta) = \alpha e^{\alpha\zeta} \int_\zeta^0 f(s)\, e^{-\alpha s} ds, \text{ where } \alpha = \frac{2\rho}{M} \quad , \tag{2.28}$$

the upper limit of the integral being chosen to satisfy (2.19). Finally the function g is found by substituting for h in (2.27).

2.6.1 Reflection and transmission of harmonic wave trains

An incident train of simple harmonic waves is obtained by taking

$$f(\zeta) = e^{ik\zeta}\ (\zeta < 0) \quad .$$

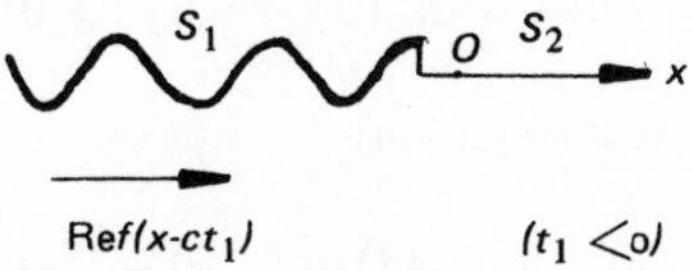

Fig. 2.8

For the string with a mass attached, (2.28) gives the transmitted wave

$$h(x - ct) = \frac{\alpha}{\alpha - ik}\, e^{ik(x-ct)} - \frac{\alpha}{\alpha - ik}\, e^{\alpha(x-ct)} \quad .$$

Concentrating momentarily on the string S_2 alone, we can say that S_2 is executing **free** motion subject to a **boundary force** exerted on it at the junction. In the terminology of section 1.4 the second term, which decays as t increases, is the **transient response** of S_2 to the arrival of the wave train. The ultimate behaviour of S_2 at any point $x > 0$ is given by the first term, which is the **steady state response.**

The motion of S_1 consists of the incident wave together with a transient and a steady state reflection. Using (2.27) we find that the steady state reflected wave is

$$\frac{ik}{\alpha - ik} e^{-ik(x+ct)} .$$

We note that the phases of the oscillations at $x = 0$ of the transmitted and reflected waves differ from the incident phase, and differ mutually by $\pi/2$. This compares with the result in section 2.5.1, where the transmitted and incident phases agree, and the reflected phase agrees when $c_2 > c_1$ but is exactly opposed to them when $c_2 < c_1$.

When we are only concerned with the steady state solution the transient parts of the reflected and transmitted waves may be omitted in the assumed forms of $g(\zeta)$ and $h(\zeta)$. For example, in section 2.6 with an incident harmonic wave form $f(\zeta) = e^{ik\zeta}$, we would assume

$$g(\zeta) = Ae^{ik\zeta} \text{ and } h(\zeta) = Be^{ik\zeta} ,$$

where A and B are complex constants to be determined from the boundary conditions at $x = 0$.

2.7 FINITE SYSTEMS

Suppose we have a system of length L. The problem of wave propagation in the system is to solve the equation

$$c^2 u_{xx} = u_{tt} \; (0 < x < L, \, t > 0)$$

with the initial conditions

$$u(x, 0) = a(x), \, u_t(x, 0) = b(x) \; (0 < x < L)$$

and given boundary conditions at $x = 0$ and $x = L$.

Whilst problems of this type are most easily dealt with by using stationary waves (Chapter 3), they may also be analysed in terms of progressive waves using the method of **successive reflection.** We shall continue to confine the discussion to cases involving **linear homogeneous boundary conditions**, which are defined as linear relations between $u(x, t)$ and its derivatives which hold at the boundaries for all values of t; examples are displayed in Figs. 2.4 and 2.7.

For the boundary $x = 0$ of the region $x < 0$ we have said that $f(x - ct)$ describes an **incident wave** if the function $f(\zeta)$ is zero for $\zeta > 0$. The corresponding reflected wave $g(x + ct)$, satisfying $g(\zeta) = 0$ for $\zeta < 0$, is then found by solving the linear relations which arise from the boundary conditions. We may therefore define a linear operator $\mathbf{r}_0$, the **right reflection operator** at O, by the relation $g(\zeta) = \mathbf{r}_0 f(\zeta)$. For example, in the case of the springy end, the result (2.16) is derived from

$$\mathbf{r}_0 f(\zeta) = f(-\zeta) - 2ke^{-k\zeta} \int_0^{\zeta} e^{ks} f(-s)\, ds \quad .$$

We now define the reflection operators at the boundaries of our given system in the following way.

Disregarding the end $x = 0$, if the wave $f(x - ct)$, where $f(\zeta) = 0$ for $\zeta > L$, is incident from the region $x < L$ on the end $x = L$, then the reflected wave will be $g(x + ct)$, where $g(\zeta) = \mathbf{r}_L\, f(\zeta)$ and $\mathbf{r}_L$ is the **right reflection operator** at $x = L$. Now disregarding the end $x = L$, if the wave $F(x + ct)$, where $F(\zeta) = 0$ for $\zeta < 0$, is incident from the region $x > 0$ on the end $x = 0$, then the reflected wave will be $G(x - ct)$, where $G(\zeta) = \mathbf{R}_0 F(\zeta)$, and $\mathbf{R}_0$ is the **left reflection operator** at $x = 0$.

For any given linear homogeneous boundary conditions the operators $\mathbf{r}_L$ and $\mathbf{R}_0$ can be found by the methods of sections 2.4 to 2.6.

We now extend $a(x)$ and $b(x)$ by taking $a(x) = b(x) = 0$ ($x < 0$ or $x > L$), and define

$$f(\zeta) = \tfrac{1}{2} a(\zeta) + \frac{1}{2c} \int_{\zeta}^{L} b(s)\, ds,$$

$$F(\zeta) = \tfrac{1}{2} a(\zeta) + \frac{1}{2c} \int_0^{\zeta} b(s)\, ds, \tag{2.29}$$

$$A = \frac{1}{2c} \int_0^{L} b(s)\, ds \quad .$$

Then $f(\zeta) = 0$ for $\zeta > L$ and $F(\zeta) = 0$ for $\zeta < 0$, and so $f(x - ct)$ and $F(x + ct)$ are waves incident at L and 0 respectively.

The solution in the absence of boundary conditions would be, from (2.9),

$$v(x, t) = -A + f(\xi) + F(\eta) \quad ,$$

where $\xi = x - ct$ and $\eta = x + ct$.

As in section 2.4, the solution $u(x, t)$ is found by adding reflected waves to $v(x, t)$. We shall first obtain a series by successive reflection at the two boundaries and then show that it is the required solution.

The wave $f(\xi)$, incident at L, is reflected as $\mathbf{r}_L f(\eta)$. This in turn undergoes reflection at 0, producing the reflected wave $\mathbf{R}_0 \mathbf{r}_L f(\xi)$, and then again at L, giving $\mathbf{r}_L \mathbf{R}_0 \mathbf{r}_L f(\eta)$, and so on. Likewise, the successive reflections of the wave $F(\eta)$ at $0, L, 0, \ldots$ are

$$\mathbf{R}_0 F(\xi),\ \mathbf{r}_L \mathbf{R}_0 F(\eta),\ \mathbf{R}_0 \mathbf{r}_L \mathbf{R}_0 F(\xi),\ \ldots\ .$$

Superposing all these waves we arrive at the series

$$u(x, t) = -A + \sum_{n=0}^{\infty} (\mathbf{R}_0 \mathbf{r}_L)^n [f(\xi) + \mathbf{R}_0 F(\xi)] + \sum_{n=0}^{\infty} (\mathbf{r}_L \mathbf{R}_0)^n\, [F(\eta) + \mathbf{r}_L f(\eta)] \tag{2.30}$$

This can be exhibited as follows

$$u(x, t) = f(\xi) + \mathbf{R}_0 F(\xi) + \mathbf{R}_0 \mathbf{r}_L f(\xi) + \mathbf{R}_0 \mathbf{r}_L \mathbf{R}_0 F(\xi) + (\mathbf{R}_0 \mathbf{r}_L)^2 f(\xi) + \ldots$$

$$-A$$

$$+ F(\eta) + \mathbf{r}_L f(\eta) + \mathbf{r}_L \mathbf{R}_0 F(\eta) + \mathbf{r}_L \mathbf{R}_0 \mathbf{r}_L f(\eta) + (\mathbf{r}_L \mathbf{R}_0)^2 F(\eta) + \ldots$$

For any value of t this is a terminating series, because the reflection of a given wave does not appear until a time L/c has elapsed since the first appearance of the wave. To show that (2.30) is the required solution we observe that

(i) each term satisfies the wave equation,

(ii) the pairs of terms joined by straight and wavy lines satisfy the boundary conditions at 0 and L respectively.

Both the wave equations and the boundary conditions are linear homogeneous. Hence the sum $u(x, t)$ also satisfies the wave equation and the boundary conditions, as required.

2.7.1 One end fixed and the other end free

Let the initial conditions be

$$u(x, 0) = a(x) \text{ and } u_t(x, 0) = b(x),\ (0 < x < L)\quad ,$$

and define $f(\zeta)$, $F(\zeta)$, A as in (2.29).

Let the boundary conditions be

$$u(0, t) = 0 \text{ and } u_x(L, t) = 0 \quad . \tag{2.31}$$

Adapting the results of sections 2.4.1 and 2.4.2 to these boundary conditions we find

$$\mathsf{R}_0 F(\zeta) = -F(-\zeta)$$

and

$$\mathsf{r}_L f(\zeta) = f(2L - \zeta) \quad .$$

Hence, from (2.30), the solution is given by

$$\begin{aligned} u(x, t) + A &= f(\xi) - F(-\xi) - f(2L + \xi) + F(-2L - \xi) + f(4L + \xi) - \ldots \\ &\quad + F(\eta) + f(2L - \eta) - F(-2L + \eta) - f(4L - \eta) + F(-4L + \eta) + \ldots \end{aligned} \tag{2.32}$$

where $\xi = x - ct$ and $\eta = x + ct$.

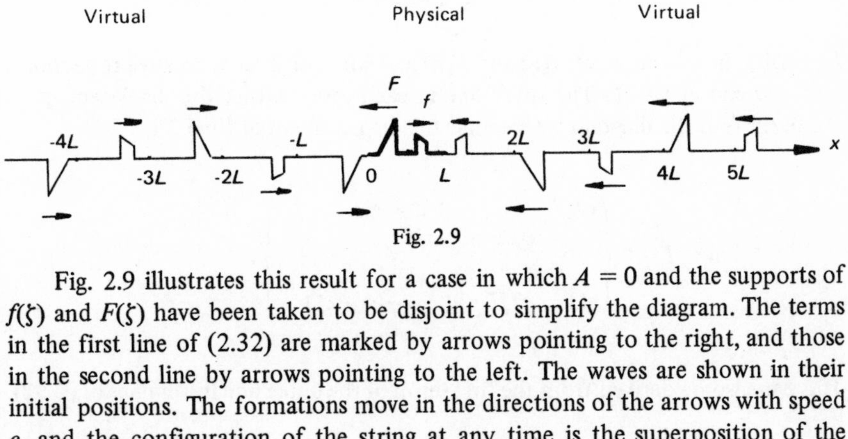

Fig. 2.9

Fig. 2.9 illustrates this result for a case in which $A = 0$ and the supports of $f(\zeta)$ and $F(\zeta)$ have been taken to be disjoint to simplify the diagram. The terms in the first line of (2.32) are marked by arrows pointing to the right, and those in the second line by arrows pointing to the left. The waves are shown in their initial positions. The formations move in the directions of the arrows with speed c, and the configuration of the string at any time is the superposition of the formations at that time in the interval $0 < x < L$. It is evident that each formation is periodic with period $4L$, and hence the motion of every point is periodic with period $4L/c$. This can be verified by substituting $t + (4L/c)$ for t in (2.32) and examining the terms of the series.

2.7.2 Example

A uniform bar of length L is fixed at the end $x = 0$ and is free at the end $x = L$. The bar is stretched to a length $L + \epsilon$, within its linear elastic limit, and released from rest at time $t = 0$. Find the longitudinal displacement of all points of the bar at a subsequent time $t = T$, where $T < L/c$.

Solution

In this case $a(x) = \epsilon x/L$ and $b(x) = 0$.

$$\text{Hence } f(\zeta) = F(\zeta) = \begin{Bmatrix} \epsilon\zeta/2L & , 0 < \zeta < L \\ 0 & , \zeta < 0 \text{ or } \zeta > L \end{Bmatrix}, \text{ and } A = 0 \quad .$$

The boundary conditions are again given by (2.31).

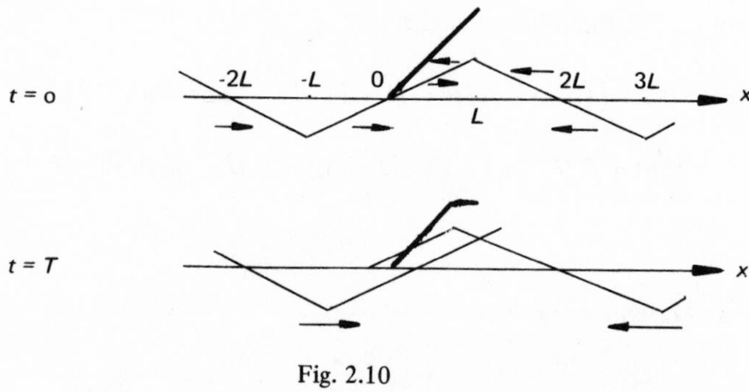

Fig. 2.10

Fig. 2.10 shows the waves $f(\xi)$ and $F(\eta)$ and some of their successive reflections, at $t = 0$ and at $t = T$. The thick line in each case denotes the displacement of the bar. From the diagram we see that the displacement at time T is

$$u(x, T) = \begin{Bmatrix} \epsilon x/L & 0 < x < L - cT \\ \\ \epsilon(1 - cT/L) & L - ct < x < L \end{Bmatrix} .$$

This can also be derived from the first terms of the upper and lower lines of (2.32).

2.8 INHOMOGENEOUS BOUNDARY CONDITIONS

When a particular point of a one-dimensional medium, say $x = x_0$, is acted on by an external force in such a way that the displacement of the point is a prescribed function of the time, the corresponding boundary condition at the point is of the form

$$u(x_0, t) = E(t)$$

where $E(t)$ is a given function. This forced motion of the point $x = x_0$ clearly affects the normal free motion of the system. What happens is that an additional wave form is superimposed on the free motion of the system and it travels through the medium at the normal wave speed c.

For example, consider a semi-infinite medium which occupies the region $x \geqslant 0$. Suppose that the end $x = 0$ is constrained to move so that

$$u(0, t) = E(t) \tag{2.33}$$

and that the initial conditions are

$$u(x, 0) = a(x), u_t(x, 0) = b(x) \; (x > 0) \quad .$$

Following the ideas of section 1.3.3, let $u = v + w$, where v and w are solutions of the wave equation and

$$\begin{array}{lll} v(x, 0) = a(x), & v_t(x, 0) = b(x), & v(0, t) = 0, \\ w(x, 0) = 0, & w_t(x, 0) = 0, & w(0, t) = E(t). \end{array}$$

Then, putting $w = p(x - ct) + q(x + ct)$, we find that $p'(\zeta)$ and $q'(\zeta)$ must be zero when $\zeta > 0$. Consequently $q(x + ct)$ is constant for $x > 0$ and $t > 0$; we may therefore discard q and write $w = p(x - ct)$, where $p(\zeta) = 0$ for $\zeta < 0$. Inserting the boundary condition we obtain

$$w(x, t) = \begin{cases} 0 \quad , & x > ct \\ E(t - x/c), & x < ct \end{cases} \quad .$$

Thus the additional disturbance produced by the forced motion at 0 travels with speed c, and its effect first reaches the point x at time x/c.

We find the function $v(x, t)$ by the method of section 2.4. It may be written

$$v(x, t) = \underset{\text{receding}}{F(x - ct)} + \underset{\text{incident}}{f(x + ct)} - \underset{\text{reflected}}{f(ct - x)} \quad ,$$

where $f(\zeta) = \frac{1}{2}a(\zeta) + \dfrac{1}{2c}\displaystyle\int_0^{\zeta} b(s)\, ds,$

$$F(\zeta) = \tfrac{1}{2}a(\zeta) - \frac{1}{2c}\int_0^{\zeta} b(s)\, ds,$$

and $a(\zeta) = b(\zeta) = 0$ for $\zeta < 0$. Hence,

for $x > ct$, $\quad u(x, t) = 0 + F(x - ct) + f(x + ct)$

$$= \tfrac{1}{2}a(x - ct) + \tfrac{1}{2}a(x + ct) + \frac{1}{2c}\int_{x-ct}^{x+ct} b(s)\, ds \quad ,$$

and for $x < ct$,

$$u(x, t) = E(t - x/c) + f(x + ct) - f(ct - x)$$
$$= E(t - x/c) + \tfrac{1}{2}a(x + ct) - \tfrac{1}{2}a(ct - x) + \frac{1}{2c}\int_{ct-x}^{ct+x} b(s)\, ds \quad .$$

2.9 THE INHOMOGENEOUS WAVE EQUATION

Forced motion of the whole system occurs when the system is subjected to a continuously distributed given external force, which may depend on both time and position. In such a case it is necessary to solve the inhomogeneous equation

$$u_{tt} = c^2 u_{xx} + p(x, t) \tag{2.34}$$

subject to some prescribed initial conditions and boundary conditions.

We now determine the effect of the forcing term $p(x, t)$ on the d'Alembert solution for an infinite medium subject to the initial conditions (2.3). We first note that the solution may be expressed as

$$u = v + w \quad , \tag{2.35}$$

where $v(x, t)$ satisfies the homogeneous wave equation (2.1) with the initial conditions (2.3) and $w(x, t)$ satisfies (2.34) with the zero initial conditions

$$w(x, 0) = 0, \quad w_t(x, 0) = 0 \quad . \tag{2.36}$$

As $v(x, t)$ is given by the d'Alembert solution (2.9) it only remains to find $w(x, t)$. We use Duhamel's Principle (see section 1.3.3), which expresses $w(x, t)$ as the superposition of a set of solutions of the homogeneous wave equation with non-zero initial conditions.

Duhamel's Principle

Let $\psi(x, t; \tau)$ be the solution of

$$\psi_{tt} = c^2 \psi_{xx}$$

satisfying $\psi(x, 0; \tau) = 0$ and $\psi_t(x, 0: \tau) = p(x, \tau)$.

Then $w(x, t) = \displaystyle\int_0^t \psi(x, t - \tau; \tau)d\tau$ satisfies

$$w(x, 0) = 0, \; w_t(x, 0) = 0$$

and

$$w_{tt} = c^2 w_{xx} + p(x, t) \quad .$$

The proof of the principle is parallel to that given in section 1.3.3 and is left as an exercise for the reader.

Now, from (2.9), we see that

$$\psi(x, t; \tau) = \frac{1}{2c} \int_{x-ct}^{x+ct} p(s, \tau) ds \quad .$$

Hence, by Duhamel's Principle,

$$\begin{aligned} w(x, t) &= \frac{1}{2c} \int_0^t \int_{x-c(t-\tau)}^{x+c(t-\tau)} p(s, \tau) ds\, d\tau \\ &= \frac{1}{2c} \iint_{\mathfrak{D}} p(s, \tau)\, ds\, d\tau \quad , \end{aligned}$$

where $\mathfrak{D}$ is the interior of the triangle with vertices at the points $P(x, t)$, $L(x - ct, 0)$ and $M(x + ct, 0)$.

Thus the solution to the inhomogeneous equation (2.34) subject to the initial conditions (2.3) is

$$\begin{aligned} u(x, t) &= \tfrac{1}{2}[a(x - ct) + a(x + ct)] + \frac{1}{2c} \int_{x-ct}^{x+ct} b(s)\, ds \\ &+ \frac{1}{2c} \iint_{\mathfrak{D}} p(s, \tau)\, ds d\tau \quad . \end{aligned} \tag{2.37}$$

The solution still depends on the values of $a(x)$ and $b(x)$ in the interval of dependence LM (see section 2.1), but in addition it also depends on the behaviour of $p(x, t)$ throughout the triangular region $\mathfrak{D}$. Consequently, for the inhomogeneous equation, the region $\mathfrak{D}$ is called the domain of dependence of the point (x, t).

PROBLEMS

2.1 Suppose $u_1(x, t)$ and $u_2(x, t)$ are the d'Alembert solutions corresponding to the initial functions $a_1(x)$, $b_1(x)$ and $a_2(x)$, $b_2(x)$ respectively. Then if

$$|a_1(x) - a_2(x)| < \epsilon, \quad |b_1(x) - b_2(x)| < \epsilon \text{ in } |x| < \infty$$

show that, for all x and for $0 \leqslant t \leqslant T$,

$$|u_1(x, t) - u_2(x, t)| < \epsilon(1 + T) \quad .$$

Deduce that the d'Alembert solution depends continuously on the initial data.

2.2 Verify the d'Alembert solution (2.9) directly by substituting it into (2.1) and (2.3).

2.3 Let C be a closed curve forming the boundary of a region S of the xt-plane in which $t > 0$. The **dependence set** of S is the union of the intervals of dependence of all points in S. Show how this set can be constructed by drawing suitable lines in the xt-plane.

2.4 Find the region of influence of the interval $x_1 < x < x_2$ (that is, the union of the regions of influence of all the points in the interval).

2.5 Use the d'Alembert formula to find the solution of the wave equation in $|x| < \infty$, $t > 0$ when

(i) $a(x) = 0$,
$b(x) = a \sin nx$, $x < 0$, (a, n constants)
$= 0$, $x > 0$.

(ii) $a(x) = h/(1 + x^2)$, h constant,
$b(x) = 0$.

(iii) $a(x) = 0$,
$b(x) = 1$, $0 < x < \alpha$, (α constant)
$= 0$, otherwise.

2.6 A wave $f(x - ct)$ travels in the string of section 2.2. Show that $\mathcal{F}(x, t) = c\mathcal{E}(x, t)$, so that the energy is transmitted with velocity c.

2.7 In the string of section 2.2 let $u = \operatorname{Re}\psi$, where $\psi = Ce^{ik(x-ct)}$ (C complex). Show that the time average of the energy flux over one period $(2\pi/kc)$ of oscillation is $\frac{1}{2}Tck^2|C|^2$.

2.8 A semi-infinite string occupies the region $x \geqslant 0$ and it is free to perform transverse vibrations, with the point $x = 0$ fixed. Given that the initial conditions are

$$u(x, 0) = a(x),\ u_t(x, 0) = b(x) \quad (x > 0)$$

show that

$$u(x, t) = \tfrac{1}{2}[a(x + ct) - a(ct - x)] + \frac{1}{2c}\int_{ct-x}^{x+ct} b(s)\,ds,\ x < ct$$

$$= \tfrac{1}{2}[a(x + ct) + a(x - ct)] + \frac{1}{2c}\int_{x-ct}^{x+ct} b(s)\,ds,\ x > ct \ .$$

2.9 Find the displacement $u(x, t)$ of the string in Problem 2.8 when the end $x = 0$ is (i) free (ii) attached to a damper of strength R.

2.10 Find the function $g(\zeta)$ for the springy end condition of section 2.4.3 in the case when

$$f(\zeta) = \begin{Bmatrix} 1, -2h < \zeta < -h \\ 0, \text{otherwise} \end{Bmatrix} \quad \text{and } k = l^{-1} \quad .$$

Draw the configurations of the string at times $5l/4c$, $7l/4c$, $3l/c$.

2.11 A string of linear density αM occupying the region $x < 0$ terminates in a smooth ring of mass M which is free to move transversely. Show that, if $f(x - ct)$ is the incident wave, then the reflected wave is $g(x + ct)$, where

$$g(\zeta) = -f(-\zeta) + 2\alpha e^{-\alpha\zeta} \int_0^\zeta e^{\alpha s} f(-s)\, ds \quad .$$

2.12 The junction of the strings in section 2.5.1 is attached to a spring which provides a transverse restoring force $T\beta$ times the displacement, and a wave $e^{ik(x-c_1t)}$ is incident in S_1. Show that the steady state reflected and transmitted waves are $Ae^{-ik(x+c_1t)}$ and $Be^{ik(nx-c_1t)}$, where $B = 1 + A = 2(n + 1 + i\beta k^{-1})^{-1}$ and $n = c_1/c_2$. Find the mean energy flux (see Problem 2.7) in all three waves and verify that the mean total energy in the strings is conserved.

2.13 An infinite uniform string stretched at tension T has a damping device at $x = 0$ exerting a force kT times the speed. A transverse wave $f(x - ct)$ approaches the origin from negative x-values. Find the reflected and transmitted waves.

2.14 Find the reflected waves in $x < 0$ if the string in section 2.5.1 is rigidly fixed at $x = L$. [Observe that there is an infinite series of reflected waves with amplitudes diminishing in geometric progression. Find the *second* wave in the series].

2.15 A uniform string extends from $x = -L$ to $x = \infty$. Its tension is T. The end $x = -L$ is attached to a light ring which is free to slide on a smooth rod transverse to the string. At $x = 0$ a light spring is attached which exerts a force λT times the deflection of the string at $x = 0$, towards the equilibrium line. A transverse wave $\cos(kx + \omega t)$ approaches the origin from $x = \infty$. Investigate the steady state and show that, for $-L < x < 0$, the transverse deflection is

$$\frac{2k \cos k(L + x) \cos(\omega t - \epsilon)}{\sqrt{[k^2 - 2k\,\lambda \sin kL\, \cos kL + \lambda^2 \cos^2 kL]}} \quad ,$$

where $\cos\epsilon$ has the sign of $k\cos kL$ and $\tan\epsilon = \tan kL - \lambda/k$.

2.16 A uniform string is fixed at its ends $x = 0$, $x = L$ and is initially at rest with displacement

$$u(x, 0) = \epsilon x^2(L - x) \quad .$$

Determine the displacement of the mid-point of the string at time $t = 5L/4c$.

2.17 A uniform string of length $2L$ is fastened to two rigid supports at the points $x = \pm L$. At time $t = 0$ the string is released from rest in a displaced configuration given by $u(x, 0) = a(x)$, where $a(x)$, defined on $[-L, L]$, is an even function. Find the period of the resulting vibrations and show that after one quarter-period the string is momentarily undeflected.

2.18 Solve the equation

$$u_{xx} - u_{tt} = xt, \quad |x| < \infty,\ t > 0$$

subject to $u(x, 0) = \alpha \sin x$, $u_t(x, 0) = \alpha \cos x$.

2.19 Solve the equation

$$u_{tt} = u_{xx} + \sin t + 2\sin x, \quad |x| < \infty,\ t > 0$$

subject to $u(x, 0) = 0$, $u_t(x, 0) = \dfrac{1}{1+x^2}$.

2.20 Solve the equation

$$u_{xx} - u_{tt} = x^2 - t^2, \quad |x| < \infty,\ t > 0$$

subject to $u(x, 0) = a\cos x$, $u_t(x, 0) = -a\sin x$.

CHAPTER 3

The Fourier Method of Solution

3.1 OUTLINE

Whilst the analytical and graphical procedures described in Chapter 2 readily provide solutions for infinite or semi-infinite systems, they become more complicated when applied to finite systems, as we have seen in section 2.7. Indeed, with conditions like the springy-end condition of section 2.4.3 the analysis often becomes intractable.

In this chapter and the next we take a different approach, which has three stages.

(i) **Separation of the variables**: we seek solutions of the wave equation which have the form $X(x)\,T(t)$.

(ii) **Normal solutions**: we select those solutions which satisfy appropriate homogeneous boundary conditions.

(iii) **Superposition**: we form a linear combination of the normal solutions which satisfies the initial conditions.

A general class of problems which can be solved in this way is discussed in Chapter 4.

The present chapter deals principally with the classical wave equation subject to the simplest boundary conditions. We shall find that the desired linear combinations of normal solutions are obtained from trigonometrical expansions called Fourier series, which form the subject of the next section.

3.2 FOURIER SERIES

Consider the **trigonometric polynomial**

$$S_N(x) = \tfrac{1}{2}a_0 + \sum_{n=1}^{N} (a_n \cos nx + b_n \sin nx) \quad .$$

$S_N(x)$ is periodic with period 2π. In general the graph of a trigonometric polynomial is much more complicated than the smooth and symmetrical sinusoidal curve. Fig. 3.1 shows the trigonometric polynomial

$$y(x) = \cos x + \tfrac{1}{2}\cos 2x + \tfrac{1}{4}\cos 3x \quad .$$

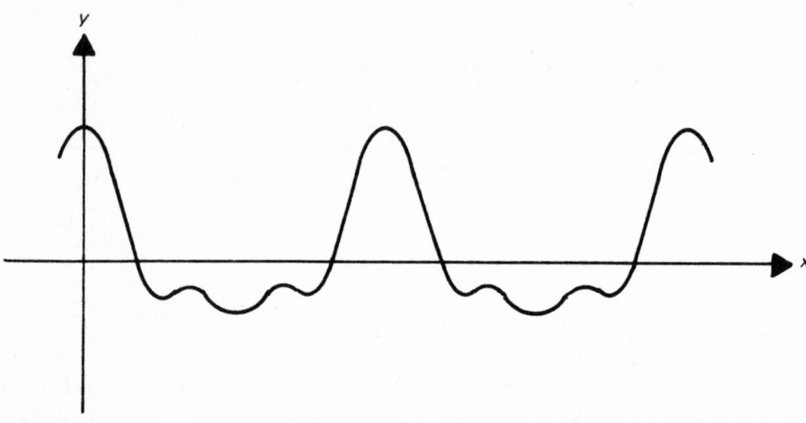

Fig. 3.1

Clearly an arbitrary function can be approximated by a trigonometric polynomial. It is natural to ask whether a sequence $S_N(x)$ $(N = 0, 1, 2, \ldots)$ can be found which converges to a given function $f(x)$ of period 2π; in other words, whether $f(x)$ can be expanded in an infinite trigonometric series with partial sums $S_N(x)$. This is, in fact, possible for a wide class of functions; sufficient conditions are given later.

3.2.1 Evaluation of the Fourier coefficients

Suppose the function $f(x)$, of period 2π, is the sum of a convergent series

$$f(x) = \tfrac{1}{2}a_0 + \sum_{n=1}^{\infty} (a_n \cos nx + b_n \sin nx) \quad . \tag{3.1}$$

This series is a linear combination of the functions

$$1, \cos x, \sin x, \cos 2x, \sin 2x, \ldots \quad . \tag{3.2}$$

The constants a_n, b_n are found by using the **mutual orthogonality** of these functions on the interval $(0, 2\pi)$, namely the easily verified property that

$$\int_0^{2\pi} \phi(x)\, \psi(x)\, dx = 0 \tag{3.3}$$

for any distinct pair (ϕ, ψ) in the set (3.2). We note also that $\int_0^{2\pi} \cos^2 nx \, dx = \pi$ and $\int_0^{2\pi} \sin^2 nx \, dx = \pi$ for $n = 1, 2, \ldots$. We now obtain expressions for $\int_0^{2\pi} f(x)\, dx$, $\int_0^{2\pi} f(x) \cos nx \, dx$ and $\int_0^{2\pi} f(x) \sin nx \, dx$ by substituting the series (3.1) for $f(x)$ and integrating term by term, assuming this to be justifiable. In each case the orthogonality annihilates all but one of the integrated terms, and we find

$$a_n = \frac{1}{\pi} \int_0^{2\pi} f(x) \cos nx \, dx, \; n = 0, 1, 2, \ldots, \tag{3.4}$$

$$b_n = \frac{1}{\pi} \int_0^{2\pi} f(x) \sin nx \, dx, \; n = 1, 2, \ldots \quad . \tag{3.5}$$

These are called the **Fourier coefficients** of $f(x)$, and they can be defined by (3.4) and (3.5) even if the series (3.1) does not converge. Whenever they exist we define the **Fourier series** of $f(x)$ by

$$f(x) \sim \tfrac{1}{2} a_0 + \sum_{n=1}^{\infty} (a_n \cos nx + b_n \sin nx) \quad . \tag{3.6}$$

The symbol $\sim$ means that the series does not necessarily converge. It can be replaced by an equality only if $f(x)$ satisfies suitable conditions.

It is clear from (3.4) and (3.5) that the Fourier coefficients of the function $f(x) \equiv 0$ are zero, and also that the Fourier series of any function is unique.

3.2.2 Conditions for convergence

Many different sets of sufficient conditions are known for the convergence of Fourier series. We shall use a simple criterion which is adequate for all vibration problems.

DEFINITION

The function $f(x)$ is said to be piecewise smooth on the interval $[a, b]$ if both $f(x)$ and $f'(x)$ are continuous on $[a, b]$ except for a finite number of jump discontinuities. A function is said to be piecewise smooth for $|x| < \infty$ if it is piecewise smooth on every finite interval.

THEOREM

The Fourier series of a piecewise smooth function $f(x)$ of period 2π converges for all values of x to the value

$$\tfrac{1}{2} [f(x + 0) + f(x - 0)] \quad , \tag{3.7}$$

where $f(x+0) = \lim_{\substack{h\to 0\\ h>0}} f(x+h)$ and $f(x-0) = \lim_{\substack{h\to 0\\ h>0}} f(x-h)$.

A proof of this theorem may be found in Tolstov (1962). At points of continuity of $f(x)$ we have

$$f(x+0) = f(x-0) = f(x) \quad ,$$

and so the Fourier series will converge to $f(x)$.

3.2.3 Example

Find the Fourier series of the function

$$\begin{aligned} f(x) &= x \quad , \quad 0 \leqslant x < \pi \\ &= \pi \quad , \quad \pi \leqslant x < 2\pi \\ f(x+2\pi) &= f(x), \ |x| < \infty \quad . \end{aligned}$$

Solution

From (3.4) we have

$$a_0 = \frac{1}{\pi}\left(\int_0^{\pi} x\,dx + \int_{\pi}^{2\pi} \pi\,dx\right) = \frac{3\pi}{2}$$

and

$$a_n = \frac{1}{\pi}\left(\int_0^{\pi} x\cos nx\,dx + \int_{\pi}^{2\pi} \pi\cos nx\,dx\right) = \frac{(-1)^n - 1}{\pi n^2} \quad , n = 1, 2, \ldots .$$

Hence $a_{2n} = 0$ and $a_{2\pi+1} = -\dfrac{2}{\pi(2n+1)^2}$.

Similarly, from (3.5) we have

$$b_n = \frac{1}{\pi}\left(\int_0^{\pi} x\sin nx\,dx + \int_{\pi}^{2\pi} \pi\sin nx\,dx\right) = -\frac{1}{n} \quad .$$

Hence $f(x) \sim \dfrac{3\pi}{4} - \dfrac{2}{\pi}\sum_{n=0}^{\infty} \dfrac{\cos(2n+1)x}{(2n+1)^2} - \sum_{n=1}^{\infty} \dfrac{\sin nx}{n}$.

Since $f(x)$ is piecewise smooth the series converges to $f(x)$ everywhere except at the points of discontinuity $x = 0, \pm 2\pi, \ldots$. At these points the series converges to $\frac{1}{2}f(+0) + \frac{1}{2}f(-0) = \frac{1}{2}(0+\pi) = \pi/2$.

3.2.4 Function defined on an interval

If $f(x)$ is defined only on the interval $(0, 2\pi)$, it is still represented on that interval by the Fourier series (3.6); the series corresponds to a periodic extension of $f(x)$ to the whole line.

If $f(x)$ is defined on the interval $(c, c + 2\pi)$, where c is constant, the periodic extension of $f(x)$ is used to evaluate a_n and b_n. A clearly equivalent method is to take c and $c + 2\pi$ as the limits of the integrals in (3.4) and (3.5).

If we wish to expand $f(x)$ in the interval $[0, 2L]$, we write $s = \pi x/L$ and $g(s) = f(sL/\pi)$. Then $g(s)$ is defined in $(0, 2\pi)$ and its Fourier series is

$$g(s) \sim \tfrac{1}{2}a_0 + \sum_{n=1}^{\infty} (a_n \cos ns + b_n \sin ns) \quad ,$$

where $a_n = \dfrac{1}{\pi}\displaystyle\int_0^{2\pi} g(s) \cos ns \, ds$ and $b_n = \dfrac{1}{\pi}\displaystyle\int_0^{2\pi} g(s) \sin ns \, ds.$

Hence $$f(x) \sim \tfrac{1}{2}a_0 + \sum_{n=1}^{\infty} \left(a_n \cos \frac{n\pi x}{L} + b_n \sin \frac{n\pi x}{L} \right) \quad , \tag{3.8}$$

where $a_n = \dfrac{1}{L}\displaystyle\int_0^{2L} f(x) \cos \frac{n\pi x}{L} \, dx$ and $b_n = \dfrac{1}{L}\displaystyle\int_0^{2L} f(x) \sin \frac{n\pi x}{L} \, dx$.

When the function $f(x)$ is defined on $(-L, L)$ it may similarly be shown that the Fourier series of $f(x)$ is again given by (3.8), where

$$a_n = \frac{1}{L}\int_{-L}^{L} f(x) \cos \frac{n\pi x}{L} \, dx, \; b_n = \frac{1}{L}\int_{-L}^{L} f(x) \sin \frac{n\pi x}{L} \, dx \quad .$$

3.2.5 Half-range Fourier series

Let f be an *even* function, defined on the interval $(-L, L)$. Then as $\cos \dfrac{n\pi x}{L}$ is also even and $\sin \dfrac{n\pi x}{L}$ is odd we see from (3.8) that the Fourier coefficients of $f(x)$ are

$$a_n = \frac{2}{L}\int_0^{L} f(x) \cos \frac{n\pi x}{L} \, dx, \; b_n = 0 \quad . \tag{3.9}$$

Thus the Fourier series of an even function contains only cosines, and

$$f(x) \sim \tfrac{1}{2}a_0 + \sum_{1}^{\infty} a_n \cos \frac{n\pi x}{L} \quad .$$

Now let f be an *odd* function defined on the interval $(-L, L)$. Then using (3.8), we find that the Fourier coefficients of $f(x)$ are

$$a_n = 0, \; b_n = \frac{2}{L}\int_0^L f(x) \sin \frac{n\pi x}{L} \quad . \tag{3.10}$$

So the Fourier series of an odd function contains only sines, and

$$f(x) \sim \sum_{n=1}^{\infty} b_n \sin \frac{n\pi x}{L} \; dx \quad .$$

A problem which often arises is that of expanding a function f defined on the interval $(0, L)$ in a cosine series or a sine series. To expand $f(x)$ in a cosine series, we proceed as follows: Make an *even* extension of $f(x)$ from the interval $(0, L)$ to the interval $(-L, L)$, so that $f(-x) = f(x)$. Then the previous results apply to the even extension of $f(x)$, so that its Fourier coefficients are given by (3.9) which only involves the values of $f(x)$ in the interval $(0, L)$.

To expand $f(x)$ in a sine series, we make the *odd* extension of $f(x)$ from the interval $(0, L)$ to the interval $(-L, L)$, so that $f(-x) = -f(x)$. Then its Fourier coefficients are given by (3.10) which again only involves the values of $f(x)$ in the interval $(0, L)$.

3.2.6 Example

Given that $f(x) = x$, $0 < x < L$, find

(a) a Fourier cosine series for $f(x)$,

(b) a Fourier sine series for $f(x)$,

valid in the interval $0 < x < L$.

Solution

(a) We make an even extension of $f(x)$, so that $f(x) = |x|$ in $(-L, L)$. Then from (3.9), $b_n = 0$ and

$$a_n = \frac{2}{L}\int_0^L x \cos \frac{n\pi x}{L} \; dx = \frac{2L}{(n\pi)^2}\,[(-1)^n - 1]\,, n \neq 0 \quad ,$$

$$a_0 = \frac{2}{L}\int_0^L x \; dx = L \quad .$$

Hence $x \sim \frac{1}{2}L - \frac{4L}{\pi^2} \sum_{n=0}^{\infty} \frac{1}{(2n+1)^2} \cos \frac{(2n+1)\pi x}{L}$.

(b) We make an odd extension of $f(x)$, so that $f(x) = x$ in $(-L, L)$. Then from (3.10), $a_n = 0$ and

$$b_n = \frac{2}{L} \int_0^L x \sin \frac{n\pi x}{L} \, dx = \frac{2L}{n\pi} (-1)^{n+1} .$$

Hence $$x \sim \frac{2L}{\pi} \sum_{n=1}^{\infty} \frac{(-1)^{n+1}}{n} \sin \frac{n\pi x}{L} .$$

3.3 EXAMPLES OF THE FOURIER METHOD

PROBLEM

Obtain the solution of the wave equation

$$c^2 u_{xx} = u_{tt} , \quad 0 < x < L, \; t > 0 \quad , \tag{3.11}$$

which satisfies the boundary conditions

$$u(0, t) = u(L, t) = 0, \; t \geqslant 0 \quad , \tag{3.12}$$

and the initial conditions

$$u(x, 0) = a(x), \; 0 \leqslant x \leqslant L \quad , \tag{3.13}$$

$$u_t(x, 0) = b(x), \; 0 \leqslant x \leqslant L \quad , \tag{3.14}$$

where $a(x)$ and $b(x)$ are given functions.

3.3.1 Separation of the variables

For a solution of the form

$$X(x) \, T(t)$$

equation (3.11) becomes

$$\frac{1}{X} \frac{d^2 X}{dx^2} = \frac{1}{c^2 T} \frac{d^2 T}{dt^2} , \qquad 0 < x < L, \; t > 0 \quad . \tag{3.15}$$

The left-hand side of this equation is a function of x only, and the right-hand side is a function of t only. Therefore (3.15) can be true throughout the required domain only if both sides have the same *constant* value, which we shall denote by $-\lambda$. This gives the pair of ordinary differential equations

$$\frac{d^2X}{dx^2} = -\lambda X \quad , \tag{3.16}$$

$$\frac{d^2T}{dt^2} = -\lambda c^2 T \quad . \tag{3.17}$$

3.3.2 The normal solutions

These are defined as functions $X(x)\,T(t)$ which satisfy the equation (3.11) and the boundary conditions (3.12). In any normal solution the function $X(x)$, called a **normal mode**, must therefore satisfy (3.16) and the conditions

$$X(0) = X(L) = 0 \quad . \tag{3.18}$$

Three possible cases now arise, namely λ positive, negative, and zero. However, when $\lambda \leqslant 0$, the only solution to (3.16) which satisfies (3.18) is $X(x) = 0$. In the case when λ is positive, say $\lambda = k^2$, equation (3.16) gives

$$X(x) = A \cos kx + B \sin kx \quad . \tag{3.19}$$

Since $X(0) = 0$, $A = 0$, and so $X(x) = B \sin kx$. The condition $X(L) = 0$ now restricts k to be a solution of

$$\sin kL = 0 \quad .$$

Hence k may take any of the values k_n, where

$$k_n = \frac{n\pi}{L} \quad , \quad n = 1, 2, \ldots \quad . \tag{3.20}$$

From (3.17) we have

$$T(t) = C \cos kct + D \sin kct \quad ,$$

where C and D are arbitrary constants. Hence, in this problem, every normal solution has the form

$$\sin k_n x \; (C \cos k_n ct + D \sin k_n ct) \quad . \tag{3.21}$$

The functions $X_n(x) = \sin k_n x$ are the normal modes and the quantities $\omega_n = k_n c$ are the normal frequencies of oscillation.

3.3.3 Superposition of normal solutions

Both the wave equation (3.11) and the boundary conditions (3.12) are linear homogeneous. Therefore any linear combination of the normal solutions (3.21), with constant coefficients, is itself a solution of (3.11) satisfying (3.12). The most general such linear combination is the infinite series

$$u(x, t) = \sum_{n=1}^{\infty} \sin k_n x \,(C_n \cos k_n ct + D_n \sin k_n ct) \quad . \tag{3.22}$$

Assuming the series to be convergent and differentiable term by term, we find, from the initial conditions (3.13) and (3.14), that

$$a(x) = \sum_{n=1}^{\infty} C_n \sin \frac{n\pi x}{L} \tag{3.23}$$

and

$$b(x) = \frac{\pi c}{L} \sum_{n=1}^{\infty} n\, D_n \sin \frac{n\pi x}{L} \quad . \tag{3.24}$$

We recognise these expressions as Fourier sine series expansions in the interval $(0, L)$. The coefficients are given by (3.10), and we obtain

$$C_n = \frac{2}{L} \int_0^L a(x) \sin \frac{n\pi x}{L}\, \mathrm{d}x \tag{3.25}$$

and

$$D_n = \frac{2}{n\pi c} \int_0^L b(x) \sin \frac{n\pi x}{L}\, \mathrm{d}x \quad . \tag{3.26}$$

Finally, substituting for C_n and D_n in (3.22), we have

$$u(x, t) = \sum_{n=1}^{\infty} \sin \frac{n\pi x}{L} \left\{ \frac{2}{L} \cos \frac{n\pi ct}{L} \int_0^L a(s) \sin \frac{n\pi s}{L}\, \mathrm{d}s \right.$$

$$\left. + \frac{2}{n\pi c} \sin \frac{n\pi ct}{L} \int_0^L b(s) \sin \frac{n\pi s}{L}\, \mathrm{d}s \right\} \quad . \tag{3.27}$$

3.3.4 Comments on the procedure

(i) Equation (3.27) is a formal solution to the problem specified by (3.11) to (3.14), and it is not necessarily a convergent series. Fortunately it can be shown

(Tolstov 1962) that whenever the problem has a solution, the series (3.27) converges to that solution.

(ii) If there are discontinuities in $a(x)$ or $b(x)$ in the interval $(0, L)$, then $u(x, t)$ cannot be continuous throughout the interval, and so there is no solution. However, a formal 'solution' can, in principle, be obtained by the d'Alembert method as discussed in section 2.1.3 together with reflections at the boundaries, and if this solution is piecewise smooth the series (3.27) will converge to it in the manner of (3.7).

(iii) The same normal modes may apply to different equations. Consider the equation

$$u_{tt} + \lambda u_t + \beta^2 u = c^2 u_{xx} \text{ subject to } u(0, t) = u(L, t) = 0 \quad ,$$

where λ and β are constants.

The equations for $X(x)$ and $T(t)$ are $X'' = -k^2 X$

and
$$T'' + \lambda T' + (\beta^2 + k^2 c^2)\, T = 0 \quad . \tag{3.28}$$

Hence the normal modes are again $X_n(x) = \sin \dfrac{n\pi x}{L}$, as in (3.21). The normal solutions, however, are modified because $T(t)$ now has to be a solution of equation (3.28).

3.3.5 Other homogeneous boundary conditions

(i) Free ends: If the conditions (3.12) are replaced by

$$u_x(0, t) = u_x(L, t) = 0 \quad ,$$

the normal modes are the functions

$$\cos k_n x, \text{ where } k_n = \frac{n\pi}{L}, \quad n = 0, 1, 2, \ldots \quad ,$$

and the solution is expanded in the series

$$u(x, t) = C_0 + D_0 t + \sum_{n=1}^{\infty} \cos k_n x (C_n \cos \omega_n t + D_n \sin \omega_n t) \quad .$$

The initial functions $a(x)$ and $b(x)$ are then expanded in cosine series.

(ii) One fixed end and one free end: The conditions are

$$u(0, t) = 0, u_x(L, t) = 0 \quad . \tag{3.29}$$

The normal modes are functions of the form $\sin kx$, where $\cos kL = 0$, giving

$$k_n = (n + \tfrac{1}{2}) \frac{\pi}{L}, \qquad n = 0, 1, 2, \ldots . \tag{3.30}$$

This case is illustrated in section 3.3.6.

(iii) Springy ends: Conditions of the type $\alpha u + \beta u_x = 0$ $(\alpha \neq 0, \beta \neq 0)$ also gives rise to normal modes of the form (3.19), but the quantities k_n are no longer integer multiples of a constant, and the ordinary Fourier series expansions cannot be used. We consider this case in Chapter 4.

3.3.6 Example

Solve

$$u_{tt} = c^2 u_{xx}, 0 < x < L, t > 0$$

subject to the boundary conditions

$$u(0, t) = 0, u_x(L, t) = 0$$

and the initial conditions

$$u(x, 0) = \alpha x/L, u_t(x, 0) = 0.$$

A corresponding physical problem is that of the torsional motion of a thin bar, which is fixed at one end and free at the other end, when the free end is twisted through an angle α and released from rest. From (3.30) the normal solutions are of the form

$$\sin k_n x(A \cos k_n ct + B \sin k_n ct), \text{ where}$$

$$k_n = (n + \tfrac{1}{2}) \pi/L, n = 0, 1, 2, \ldots \quad . \tag{3.31}$$

The general superposition of normal solutions is

$$u(x, t) = \sum_{n=0}^{\infty} \sin \frac{(2n+1)\pi x}{2L} \left\{ A_n \cos \frac{(2n+1)\pi ct}{2L} + B_n \sin \frac{(2n+1)\pi ct}{2L} \right\}.$$

The initial conditions require that

$$\frac{\alpha x}{L} = \sum_{n=0}^{\infty} A_n \sin \frac{(2n+1)\pi x}{2L}$$

and

$$0 = \sum_{n=0}^{\infty} B_n \frac{(2n+1)\pi c}{2L} \sin \frac{(2n+1)\pi x}{2L} .$$

Hence $B_n = 0$.

To show that $a(x) = \alpha x/L$ can be expanded in a series of sines of odd multiples of $\pi x/2L$ we extend $a(x)$ to the domain $(0, 2L)$ by defining

$$a(x) = (2 - x/L), < x < 2L \quad ,$$

as shown in Fig. 3.2.

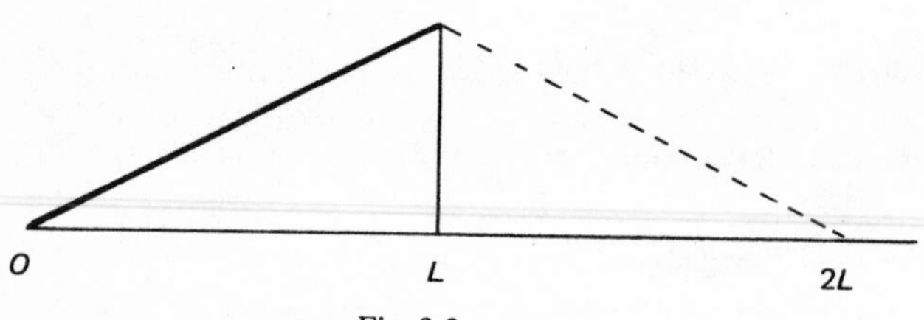

Fig. 3.2

In the expansion of $a(x)$ in a sine series in the interval $(0, 2L)$, all the even coefficients are zero. The expansion therefore has the desired form. Adapting (3.10) we obtain

$$A_n = \frac{2}{2L} \int_0^{2L} a(x) \sin \frac{(2n+1)\pi x}{2L} \, dx$$

$$= \frac{2}{L} \int_0^{L} \frac{\alpha x}{L} \sin \frac{(2n+1)\pi x}{2L} \, dx \qquad \text{by symmetry}$$

$$= \frac{(-1)^n 8\alpha}{(2n+1)^2 \pi^2} .$$

Hence $u(x, t) = \dfrac{8\alpha}{\pi^2} \displaystyle\sum_{n=0}^{\infty} \frac{(-1)^n}{(2n+1)^2} \sin \frac{(2n+1)\pi x}{2L} \cos \frac{(2n+1)\pi c t}{2L} .$

3.4 THE INHOMOGENEOUS EQUATION

3.4.1 Expansion in normal modes of the homogeneous equation

Consider the equation

$$u_{tt} + \lambda u_t - c^2 u_{xx} = p(x, t), \qquad 0 < x < L,\ t > 0 \quad . \tag{3.32}$$

This corresponds to the problem of a string subject to a constant linear damping λ and an external force density $p(x, t)$. Suppose we require a solution which satisfies given free or fixed end conditions and initial conditions

$$u(x, 0) = a(x), u_t(x, 0) = b(x) \quad .$$

Let the functions $X_n(x)$ be the normal modes of the corresponding homogeneous equation

$$u_{tt} + \lambda u_t - c^2 u_{xx} = 0 \quad . \tag{3.33}$$

Then $X_n''(x) = -k_n^2 X_n(x)$, where the values of k_n are determined by the boundary conditions.

Suppose the required solution can be expanded in the infinite series of normal modes

$$u(x, t) = \sum_{n=1}^{\infty} S_n(t) X_n(x) \quad . \tag{3.34}$$

We have seen that, for free or fixed end conditions, this is a Fourier series with coefficients $S_n(t)$, and so the expansion is valid if the solution $u(x, t)$ exists.

Let $$p(x, t) = \sum_{n=1}^{\infty} P_n(t) X_n(x) \quad .$$

Then equation (3.32) becomes

$$\sum_{n=1}^{\infty} X_n(x) \ [S_n''(t) + \lambda S_n'(t) + k_n^2 c^2 S_n(t) - P_n(t)] = 0 \quad .$$

But the Fourier coefficients of an identically zero function must vanish. Hence

$$S''_n(t) + \lambda S_n'(t) + k_n^2 c^2 S_n(t) = P_n(t),\ n = 1, 2 \ldots \tag{3.35}$$

These ordinary differential equations are easily solved, and it only remains to find the set of initial values of $S_n(0)$ and $S_n'(0)$ which enable the initial conditions to be satisfied. To do this we expand both $a(x)$ and $b(x)$ in Fourier series of the normal modes $X_n(x)$ so that

$$u(x, 0) = a(x) = \sum_{n=1}^{\infty} \alpha_n X_n(x) \quad , \tag{3.36}$$

$$u_t(x, 0) = b(x) = \sum_{n=1}^{\infty} \beta_n X_n(x) \quad . \tag{3.37}$$

Now using the series expansion (3.34) we obtain

$$\sum_{n=1}^{\infty} X_n(x) \left[S_n(0) - \alpha_n\right] = 0 \quad ,$$

$$\sum_{n=1}^{\infty} X_n(x) \left[S_n'(0) - \beta_n\right] = 0 \quad .$$

Again the Fourier coefficients must vanish, and we see that the initial conditions are satisfied when the equations (3.35) are solved subject to the initial values

$$S_n(0) = \alpha_n \text{ and } S_n'(0) = \beta_n,\ n = 1, 2, \ldots \quad .$$

3.4.2 Example
Solve the equation

$$u_{tt} + u_t - u_{xx} = \sin \frac{\pi x}{L}, 0 < x < L, t > 0 \quad , \tag{3.38}$$

with the boundary conditions $u(0, t) = u(L, t) = 0$ and the initial conditions $u(x, 0) = u_t(x, 0) = 0$.

Solution
The normal modes of the homogeneous equation $u_{tt} + u_t = u_{xx}$ with the same boundary conditions are easily found to be

$$X_n(x) = \sin \frac{n\pi x}{L}, \qquad n = 1,2, \ldots \; .$$

Consequently we look for a solution of the form

$$u(x, t) = \sum_{n=1}^{\infty} \sin \frac{n\pi x}{L} S_n(t) \quad .$$

This satisfies the given boundary conditions. Substituting for $u(x, t)$ in (3.38) and noting that the R.H.S. of (3.38) is already expressed as a series (with only one term) in the functions $X_n(x)$, we obtain

$$S_n''(t) + S_n'(t) + (n\pi/L)^2 S_n(t) = 1, \qquad n = 1$$
$$= 0, \qquad n \neq 1 \quad .$$

The initial conditions are satisfied if $S_n(0) = 0$ and $S_n'(0) = 0$ ($n = 1, 2, \ldots$). Hence $S_n(t) = 0$ ($n = 2, 3, \ldots$), and

$$S_1(t) = \left(\frac{L}{\pi}\right)^2 \left\{1 - e^{-t/2}\left(\cos qt + \frac{1}{2q}\sin qt\right)\right\} \quad ,$$

where $$q = \sqrt{\left(\frac{\pi^2}{L^2} - \frac{1}{4}\right)} \; .$$

Hence $$u(x, t) = \left(\frac{L}{\pi}\right)^2 \sin\frac{\pi x}{L}\left\{1 - e^{-t/2}\left(\cos qt + \frac{1}{2q}\sin qt\right)\right\} .$$

3.5 INHOMOGENEOUS BOUNDARY CONDITIONS

We now consider the inhomogeneous equation

$$u_{tt} + \lambda u_t - c^2 u_{xx} = p(x, t), \qquad 0 < x < L,\ t > 0, \tag{3.39}$$

subject to the initial conditions

$$u(x, 0) = a(x),\ u_t(x, 0) = b(x) \tag{3.40}$$

and the inhomogeneous boundary conditions

$$\alpha_0 u(0, t) - \beta_0 u_x(0, t) = h_0(t),\ \alpha_1 u(L, t) + \beta_1 u_x(L, t) = h_1(t) \quad , \tag{3.41}$$

where $\alpha_0, \beta_0, \alpha_1, \beta_1$ are constants and $h_0(t), h_1(t)$ are given functions.

We can reduce the problem to one with homogeneous boundary conditions by introducing an auxiliary function $w(x, t)$ which may be any C^2 function satisfying the boundary conditions (3.41). It is not necessary for w to satisfy either the homogeneous or the inhomogeneous equation; the *only* criterion is that it must be C^2 and satisfy (3.41).

When a suitable function $w(x, t)$ has been found we look for a solution of the form

$$u(x, t) = v(x, t) + w(x, t) \tag{3.42}$$

where $v(x, t)$ is a new unknown function.

The relations (3.39) to (3.41) then become

$$v_{tt} + \lambda v_t - c^2 v_{xx} = p(x, t) - w_{tt} - \lambda w_t + c^2 w_{xx} \quad , \tag{3.43}$$

$$v(x, 0) = a(x) - w(x, 0), v_t(x, 0) = b(x) - w_t(x, 0) \quad , \tag{3.44}$$

and $\alpha_0 v(0, t) - \beta_0 v_x(0, t) = 0, \alpha_1 v(L, t) + \beta_1 v_x(L, t) = 0 \quad .$ (3.45)

Thus $v(x, t)$ satisfies a new inhomogeneous equation (3.43) with the *homogeneous* boundary conditions (3.45), and its solution (in the simple cases where $\alpha_0 \beta_0 = \alpha_1 \beta_1 = 0$) may be found by the procedure of section 3.4. For general values of $\alpha_0, \beta_0, \alpha_1, \beta_1$, the methods of Chapter 4 are needed.

3.5.1 Example

Solve $u_{tt} = c^2 u_{xx}, \quad 0 < x < L, t > 0 \quad ,$ (3.46)

with boundary conditions

$$u(0, t) = 0, u(L, t) = \sin \omega t \quad , \tag{3.47}$$

and initial conditions

$$u(x, 0) = 0, u_t(x, 0) = 0 \quad .$$

Clearly $w = \dfrac{x}{L} \sin\omega t$ (3.48)

is one function which satisfies the boundary conditions.

However, in this particular problem the boundary forcing function $\sin\omega t$ fortuitously forms part of a separated solution of (3.46), and so we can simplify the work by choosing w to satisfy (3.46) as well as the boundary conditions.

Let $w = W(x) \sin kct$, where $kc = \omega$.

Then we require

$$W'' + k^2 W = 0, \text{ and } W(0) = 0, W(L) = 1 \quad .$$

Consequently $W(x) = \dfrac{\sin kx}{\sin kL} \quad ,$

giving $w(x, t) = \dfrac{\sin kx \sin kct}{\sin kL} \quad .$

The advantage of this choice of $w(x, t)$ is that, if $u(x, t) = v(x, t) + w(x, t)$, then $v(x, t)$ must satisfy the equation

$$v_{tt} = c^2 v_{xx}, 0 < x < L, t > 0$$

with boundary conditions

$$v(0, t) = 0, v(L, t) = 0$$

and initial conditions

$$v(x, 0) = - w(x, 0) = 0 \quad ,$$

$$v_t(x, 0) = - w_t(x, 0) = \frac{-kc \sin kx}{\sin kL} \quad .$$

This is a special case of the example of the Fourier method discussed in section 3.3. The final result is

$$u(x, t) = \frac{\sin(\omega x/c)}{\sin(\omega L/c)} \sin \omega t + \frac{2c\omega}{L} \sum_{n=1}^{\infty} \frac{(-1)^{n+1}}{\omega^2 - (n\pi c/L)^2} \sin \frac{n\pi x}{L} \sin \frac{n\pi ct}{L} \tag{3.49}$$

The first term can be regarded as a forced oscillation at the driving frequency ω. The series is a superposition of the normal oscillations with frequencies $n\pi c/L$, $n = 1,2, \ldots$. When $\omega = n\pi c/L$, resonance occurs, and the expression (3.49) ceases to be valid. The solution can then be obtained either by finding the limit of the expression as $\omega \to n\pi c/L$, or by beginning with a different choice of $w(x, t)$, such as (3.48).

3.6 FOURIER INTEGRALS

3.6.1 Fourier's integral theorem

We have seen that a function $f(x)$ which is piecewise smooth in $(-L, L)$ can be expressed as a superposition of harmonic functions in the form

$$\tfrac{1}{2}[f(x + 0) + f(x - 0)] = \tfrac{1}{2}a_0 + \sum_{n=1}^{\infty} \left(a_n \cos \frac{n\pi x}{L} + b_n \sin \frac{n\pi x}{L}\right), -L \leqslant x \leqslant L\,,$$

where

$$a_n = \frac{1}{L} \int_{-L}^{L} f(s) \cos \frac{n\pi s}{L} ds \text{ and } b_n = \frac{1}{L} \int_{-L}^{L} f(s) \sin \frac{n\pi s}{L} ds \quad .$$

Consequently, using the ~ symbol of section 3.2, we have

$$f(x) \sim \frac{1}{2L}\int_{-L}^{L} f(s)\,ds + \sum_{n=1}^{\infty} \frac{1}{L}\int_{-L}^{L} f(s)\cos\frac{n\pi(s-x)}{L}\,ds \quad . \tag{3.50}$$

We now consider the problem of representing a function $f(x)$ defined *on the whole real line* as a superposition of harmonic functions. Write $L = \pi/h$ and let $h \to 0$ in (3.50). Then $L \to \infty$, and the first integral vanishes provided that $\int_{-\infty}^{\infty} f(s)\,ds$ exists. We obtain

$$f(x) \sim \lim_{h\to 0} \frac{1}{\pi}\sum_{n=1}^{\infty} h \int_{-\pi/h}^{\pi/h} f(s)\cos nh(s-x)\,ds.$$

Treating the limit of the sum as an integral, we are led to anticipate the result

$$f(x) \sim \frac{1}{\pi}\int_{0}^{\infty} d\lambda \int_{-\infty}^{\infty} f(s)\cos\lambda(s-x)\,ds,$$

which is known as **Fourier's integral formula.** Of course the above argument is not a proof, but the following theorem can be proved (Tolstov 1962).

Fourier's Integral Theorem: If $f(x)$ is absolutely integrable and piecewise smooth for $|x| < \infty$, then

$$\tfrac{1}{2}[f(x+0) + f(x-0)] = \frac{1}{\pi}\int_{0}^{\infty} d\lambda \int_{-\infty}^{\infty} f(s)\cos\lambda(s-x)\,ds \quad . \tag{3.51}$$

For convenience we shall henceforward write simply $f(x)$ instead of the left-hand side of this equation.

3.6.2 Fourier transforms

The Fourier integral formula may be written in complex form by noting that, assuming the convergence of the integrals,

$$\frac{1}{2\pi}\int_{-\infty}^{\infty} d\lambda \int_{-\infty}^{\infty} f(s)e^{i\lambda(s-x)}ds = \frac{1}{\pi}\int_{0}^{\infty} d\lambda \int_{-\infty}^{\infty} \cos\lambda(s-x)\,ds = f(x) \quad , \tag{3.52}$$

because $\int_{-\infty}^{\infty} f(s) \begin{smallmatrix}\cos\\ \sin\end{smallmatrix} \lambda(s-x)\,ds$ is an $\begin{smallmatrix}\text{even}\\ \text{odd}\end{smallmatrix}$ function of λ. We define the **Fourier**

transform $F(\lambda)$ of a function $f(x)$ by

$$F(\lambda) = \int_{-\infty}^{\infty} f(s)\, e^{i\lambda s} ds \quad . \tag{3.53}$$

Then, from (3.52) we see that the inverse transform is

$$f(x) = \frac{1}{2\pi} \int_{-\infty}^{\infty} F(\lambda) e^{-i\lambda x} d\lambda \quad .$$

This displays $f(x)$ as a superposition of harmonic functions, the relative amplitudes and phases of the constituents being given by $|F(\lambda)|$ and $-\arg F(\lambda)$.

For a function $f(x)$ defined on $x \geqslant 0$ the analogues of the half-range coefficient formulae (3.9) and (3.10) are the

Fourier cosine transform $C(\lambda) = \displaystyle\int_0^{\infty} f(s) \cos \lambda s \; ds, \; \lambda \geqslant 0$ (3.54)

and the

Fourier sine transform $S(\lambda) = \displaystyle\int_0^{\infty} f(s) \sin \lambda s \; ds, \; \lambda \geqslant 0$. (3.55)

By extending $f(x)$ as an even/odd function defined on the whole real line we see from (3.51) that the inverse transforms, that is, the expansions of $f(x)$ as cosine/sine integrals, are

$$f(x) = \frac{2}{\pi} \int_0^{\infty} C(\lambda) \cos \lambda x \; d\lambda, \quad x \geqslant 0 \tag{3.56}$$

and

$$f(x) = \frac{2}{\pi} \int_0^{\infty} S(\lambda) \sin \lambda x \; d\lambda, \quad x \geqslant 0 \quad . \tag{3.57}$$

3.6.3 An example of the use of harmonic analysis

In section 2.6.1 we saw that the transmitted and reflected waves produced by an isolated load on a string were much easier to find for a harmonic incident wave $e^{ik(x-ct)}$ than for the general case. Suppose the incident wave is the square wave given by

$$\begin{aligned} f(\zeta) &= 1, \quad -l < \zeta < 0 \quad , \\ &= 0 \quad \text{otherwise} \quad . \end{aligned}$$

We first analyse $f(\zeta)$ by determining its Fourier transform

$$F(k) = \int_{-\infty}^{\infty} f(s)\, e^{iks} \mathrm{d}s = \int_{-l}^{0} e^{iks} \mathrm{d}s = \frac{2}{k} \sin\frac{kl}{2}\, e^{-ikl/2} .$$

Then

$$f(\zeta) = \frac{1}{\pi} \int_{-\infty}^{\infty} \frac{1}{k} \sin\frac{kl}{2}\, e^{-ikl/2}\, e^{-ik\zeta} \mathrm{d}k \quad . \tag{3.58}$$

Using the method of section 2.6.1 for each constituent $e^{-ik\zeta}$ in (3.58), and simplifying, we find the following expression for the reflected wave:

$$g(x + ct) = -\frac{i}{\pi} \int_{-\infty}^{\infty} \frac{1}{\alpha + ik} \sin\frac{kl}{2}\, e^{-ikl/2}\, e^{ik(x+ct)}\, \mathrm{d}k \quad .$$

This form of the result is useful when the frequency content of the reflected wave is of more importance than the wave form itself, as is the case in many applications.

3.6.4 Fourier transform of a partial differential equation

Another use of the Fourier transform is to facilitate the solution of linear partial differential equations. We illustrate the procedure in the following derivation of the d'Alembert solution.

We wish to solve

$$c^2 u_{xx} = u_{tt},\ |x| < \infty, t > 0$$

subject to the initial conditions $u(x, 0) = a(x)$, $u_t(x, 0) = b(x)$.
Suppose $u(x, t)$ has a Fourier transform

$$U(\lambda, t) = \int_{-\infty}^{\infty} u(x, t)\, e^{i\lambda x}\, \mathrm{d}x \quad .$$

Taking the Fourier transform of both sides of the equation, we obtain

$$-c^2 \lambda^2 U(\lambda, t) = U_{tt}(\lambda, t) \quad .$$

We have thus reduced the problem to the solution of an ordinary differential equation; the second variable λ enters only algebraically.

The initial conditions become

$$U(\lambda, 0) = A(\lambda),\ U_t(\lambda, 0) = B(\lambda) \quad ,$$

where $A(\lambda)$, $B(\lambda)$ are the transforms of $a(x)$, $b(x)$.

Hence

$$U(\lambda, t) = A(\lambda) \cos\lambda ct + \frac{1}{\lambda c} B(\lambda) \sin\lambda ct \tag{3.59}$$

and the solution $u(x, t)$ is obtained by finding the inverse transform of $U(\lambda, t)$.

To find the inverse transform of $A(\lambda) \cos \lambda ct$ we put

$$\cos \lambda ct = \tfrac{1}{2}(e^{i\lambda ct} + e^{-i\lambda ct})$$

and use the fact that the transform of $a(x - l)$ is $e^{i\lambda l}A(\lambda)$, where l is a constant. It then follows that the inverse transform of $A(\lambda) \cos\lambda ct$ is $\frac{1}{2}[a(x - ct) + a(x + ct)]$.

The inverse transform of $(\lambda c)^{-1}B(\lambda) \sin\lambda ct$ is

$$\frac{1}{2\pi}\int_{-\infty}^{\infty} \frac{1}{2i\lambda c} B(\lambda)\,(e^{i\lambda ct} - e^{-i\lambda ct})\, e^{-i\lambda x} d\lambda$$

$$= \frac{1}{2c}\,\frac{1}{2\pi}\int_{-\infty}^{\infty} B(\lambda)\left(\int_{x-ct}^{x+ct} e^{-i\lambda s}\, ds\right) d\lambda$$

$$= \frac{1}{2c}\int_{x-ct}^{x+ct} b(s)\, ds \quad .$$

Hence from inverting (3.59), we obtain

$$u(x, t) = \tfrac{1}{2}[a(x - ct) + a(x + ct)] + \frac{1}{2c}\int_{x-ct}^{x+ct} b(s)\, ds \quad .$$

The validity of the derivation (but not the result, as we know), is restricted to cases in which $u(x, t)$, and therefore $a(x)$ and $b(x)$, are absolutely integrable in the infinite domain. Although clearly not the best method for this problem, the Fourier transform procedure is widely used in more difficult problems to which direct integration methods cannot be applied.

PROBLEMS

3.1 Show that

(i) $1 = \frac{4}{\pi}\sum_{0}^{\infty} \frac{1}{(2n+1)} \sin(2n+1)x,\ 0 < x < n \quad .$

(ii) $x^2 = \frac{\pi^2}{3} + 4\sum_{1}^{\infty} \frac{(-1)^n}{n^2} \cos nx,\ -\pi < x < \pi \quad .$

(iii) $e^x = \dfrac{1}{\pi}\sinh\pi + \dfrac{2}{\pi}\sinh\pi \sum_1^\infty \dfrac{(-1)^n}{1+n^2}(\cos nx - n\sin nx), -\pi < x < \pi$.

(iv) $1 - x = \dfrac{2}{\pi}\sum_1^\infty \dfrac{1}{n}[1+3(-1)^n]\sin\dfrac{n\pi x}{4}, 0 < x < 4$.

(v) $\dfrac{k}{2} - \dfrac{4k}{\pi^2}\sum_0^\infty \dfrac{1}{(2n+1)^2}\cos\dfrac{2(2n+1)\pi x}{L} = \begin{cases} 2kx/L, 0 < x < L/2, \\ 2k(L-x)/L, L/2 < x < L. \end{cases}$

3.2 A uniform string of line density ρ is stretched to a tension ρc^2 and is fixed at the ends $x = 0$ and $x = L$. The point $x = b$ is drawn aside a small distance h and the string is released from rest at time $t = 0$. Show that at any subsequent time t the transverse displacement of the string $u(x, t)$ is given by

$$u(x, t) = \frac{2hL^2}{\pi^2 b(L-b)} \sum_{n=1}^\infty \frac{1}{n^2} \sin\frac{n\pi b}{L} \sin\frac{n\pi x}{L} \cos\frac{n\pi ct}{L} .$$

3.3 Find the motion of the string in Problem 3.2 if it is released from rest in the configuration

$$u(x, 0) = a\sin(\pi x/L) + \tfrac{1}{2}a\sin(2\pi x/L) .$$

3.4 An elastic bar of natural length L is stretched by an amount λL and is at rest. Thus the initial longitudinal displacements of its sections are proportional to the distance from the end $x = 0$. At time $t = 0$ both ends are released and remain free. Show that subsequent longitudinal displacements are given by

$$u(x, t) = \tfrac{1}{2}\lambda L - \frac{4\lambda L}{\pi^2}\sum_{n=1}^\infty \frac{1}{(2n-1)^2}\cos\frac{(2n-1)\pi x}{L}\cos\frac{(2n-1)\pi ct}{L} ,$$

where c is the speed of wave propagation in the bar.

3.5 Find the subsequent longitudinal displacement of the bar in Problem 3.4 if the end $x = 0$ remains fixed whilst the end $x = L$ is free.

3.6 A uniform string of length $3L$ has line density ρ and tension T. The ends of the string are fixed and two equal and opposite transverse impulses of magnitude I are applied to the points of trisection of the string. Derive an expression for the displacement of the string at any subsequent time.

3.7 Solve the equation

$$u_{xx} = u_{tt} - \sin t, 0 < x < L, t > 0,$$

subject to the boundary conditions $u_x(0, t) = u(L, t) = 0$, and the initial conditions $u(x, 0) = u_t(x, 0) = 0$.

3.8 A uniform string of length L, line density ρ and tension $c^2\rho$ is vibrating in a resisting medium such that the equation governing damped vibrations is

$$c^2 u_{xx} = u_{tt} + 2\lambda u_t \quad .$$

The end $x = 0$ is fixed, while the end $x = L$ is made to move so that its displacement is $a\cos(\pi ct/L)$. Prove that, provided that $\lambda L \ll c$, the forced oscillation (see (3.49)) is given by

$$u(x, t) = a\operatorname{cosech}\frac{\lambda L}{c}\left(\sin\frac{\pi x}{L}\cosh\frac{\lambda x}{c}\sin\frac{\pi ct}{L} - \cos\frac{\pi x}{L}\sinh\frac{\lambda x}{c}\cos\frac{\pi ct}{L}\right) \quad .$$

3.9 A string is stretched between two points distance $2L$ apart on a frame, and the frame is caused to oscillate at right angles to the string with small displacement $a \cos \omega t$. Prove that the forced oscillation of the string at distance x from its mid-point is

$$a \sec(\omega L/c) \cos(\omega x/c) \cos \omega t,$$

provided that $\omega L(c\pi)^{-1}$ is not an integer.

3.10 Use the Fourier theorem to show that

$$e^{-|x|} = \frac{2}{\pi}\int_0^\infty \frac{\cos\lambda x}{1+\lambda^2}\,d\lambda, \quad |x| < \infty \quad .$$

3.11 Use the Fourier sine transform to show that when a semi-infinite string with one end fixed at the origin is stretched along the positive half of the x-axis, and is released from rest from a position $u(x, 0) = a(x)$, $x \geqslant 0$, then

$$u(x, t) = \frac{2}{\pi}\int_0^\infty \cos\lambda ct \sin\lambda x \int_0^\infty a(s) \sin\lambda s \, ds \, d\lambda \quad .$$

3.12 Show that the Fourier series for a function $f(x)$ of period 2π may be developed in the form

$$f(x) = \sum_{n=-\infty}^{\infty} c_n e^{inx} \quad ,$$

where

$$c_n = \frac{1}{2\pi}\int_{-\pi}^{\pi} e^{-inx} f(x)\,dx,$$

and find the coefficients a_n, b_n of (3.1) in terms of c_n, c_{-n}, for $n = 0, 1, 2, \ldots$.

CHAPTER 4

The Generalised Fourier Method

4.1 NON-UNIFORM STRINGS

In this chapter an extension of the Fourier method is applied to a wider class of equations and boundary conditions. Consider, for example, the small transverse motion of a string of length L with variable density $\rho(x)$ and tension $T(x)$. The equation of motion (derived in section 1.10.3) is

$$\rho u_{tt} = (T u_x)_x, \qquad 0 < x < L, \; t > 0 \quad . \tag{4.1}$$

Suppose we have homogeneous boundary conditions of the form

$$\alpha_0 u(0, t) - \beta_0 u_x(0, t) = 0, \quad \alpha_1 u(L, t) + \beta_1 u_x(L, t) = 0 \quad , \tag{4.2}$$

where α_0, α_1, β_0, β_1, are constants. Separating the variables we find solutions of the form

$$X(x)\,(C \cos \omega t + D \sin \omega t) \quad ,$$

where the normal mode $X(x)$ must satisfy the equation

$$(TX')' + \omega^2 \rho\, X = 0$$

with $\qquad \alpha_0 X(0) - \beta_0 X'(0) = 0, \quad \alpha_1 X(L) + \beta_1 X'(L) = 0 \quad .$

As before, we require the values of ω for which a solution exists, together with the corresponding normal modes. This is an example of the **Sturm-Liouville problem**, which we shall now discuss.

4.2 THE STURM-LIOUVILLE PROBLEM

The general Sturm-Liouville problem is concerned with the solutions $X(x)$ of the equation

$$(pX')' - qX + \lambda r X = 0 \quad , \tag{4.3}$$

in an interval, where $p(x) > 0$, $r(x) > 0$, and $p'(x)$, $q(x)$ and $r(x)$ are real and continuous in the interval, subject to certain types of linear homogeneous boundary conditions.

An **eigenvalue** is any value of λ for which a not identically zero solution $X(x)$ of the equation and boundary conditions exists; the solution $X(x)$ is called an **eigenfunction** belonging to the eigenvalue λ. An eigenvalue may be **multiple**, possessing several linearly independent eigenfunctions, or **simple**, possessing only one.

In the **regular Sturm-Liouville problem** for the interval $(0, L)$ the boundary conditions are

$$\alpha_0 X(0) - \beta_0 X'(0) = 0, \alpha_1 X(L) + \beta_1 X'(L) = 0 \quad , \tag{4.4}$$

where $\alpha_0, \beta_0, \alpha_1, \beta_1$ are real constants.

4.2.1 The fundamental expansion theorem

THEOREM

(a) Every regular Sturm-Liouville system has an infinite set of eigenvalues $\lambda_1, \lambda_2, \ldots$.

(b) Each eigenvalue λ_n is simple and real, and possesses a real eigenfunction $X_n(x)$.

(c) The functions $\{r^{1/2} X_n\}$ are mutually orthogonal in the interval.

(d) The eigenfunctions $\{X_n(x)\}$ form a **complete set**, which means that any piecewise continuous function $f(x)$ defined on the interval can be expanded in the form

$$\sum_{n=1}^{\infty} F_n X_n(x) \quad , \tag{4.5}$$

the series converging in the mean to $f(x)$, that is,

$$\lim_{N \to \infty} \int_0^L r(x) \left\{ \sum_{n=1}^{N} F_n X_n(x) - f(x) \right\}^2 \mathrm{d}x = 0 \ .$$

(e) The coefficients F_n are given by

$$F_n \int_0^L r X_n^2 \, \mathrm{d}x = \int_0^L r f X_n \, \mathrm{d}x \quad . \tag{4.6}$$

Formula (4.6) follows from the orthogonality property of the eigenfunctions if the validity of term-by-term integration is assumed, but the complete proof of the theorem is difficult, and there are many treatments (see, for example, Courant and Hilbert (1953)). The results (b) and (c) are easy to establish, and this will be done in section 4.2.4. First we consider an example.

4.2.2 Application to the classical wave equation

In seeking the normal solutions of the equation $c^2u_{xx} = u_{tt}$ subject to the boundary conditions (4.2) we find that normal modes $X(x)$ must satisfy the equation $X'' + \lambda X = 0$, which is equation (4.3) with $p = 1, q = 0, r = 1$, subject to the conditions (4.4). We have already considered the case $\beta_0 = \beta_1 = 0$ in section 3.3 and the case $\beta_0 = 0$, $\alpha_1 = 0$ in section 3.4.1. The Fourier expansions there obtained are now seen as special applications of the fundamental theorem. We now apply the theorem to an example with one fixed and one springy end.

Problem

Solve

$$c^2 u_{xx} = u_{tt}, 0 < x < L, t > 0 \tag{4.7}$$

subject to

$$u(0, t) = 0, u_x(L, t) + \alpha u(L, t) = 0, t > 0 \tag{4.8}$$

and

$$u(x, 0) = a(x), u_t(x, 0) = b(x), 0 \leqslant x \leqslant L \quad . \tag{4.9}$$

Solution

Following the usual procedure we find that the normal solutions are of the form

$$X(x)\,(C \cos kct + D \sin kct) \quad ,$$

where $X'' + k^2X = 0, X(0) = 0, X'(L) + \alpha X(L) = 0$.

Hence $X(x) = \sin kx$

and

$$k \cos kL + \alpha \sin kL = 0 \quad . \tag{4.10}$$

Let the roots of this equation be $k_1, k_2, \ldots$ (see Fig. 4.1). Then the eigenvalues $\{\lambda_n\}$ and the eigenfunctions (normal modes) $\{X_n(x)\}$ of the problem are given by

$$\lambda_n = k_n^2 \text{ and } X_n(x) = \sin k_n x, n = 1, 2, \ldots \ .$$

From the theorem we know that the eigenfunctions are orthogonal and form a complete set, and using (4.10) we find that

$$\int_0^L \sin^2 k_n x \, dx = \tfrac{1}{2}\left(L + \frac{\alpha}{\alpha^2 + k_n{}^2}\right) \quad . \tag{4.11}$$

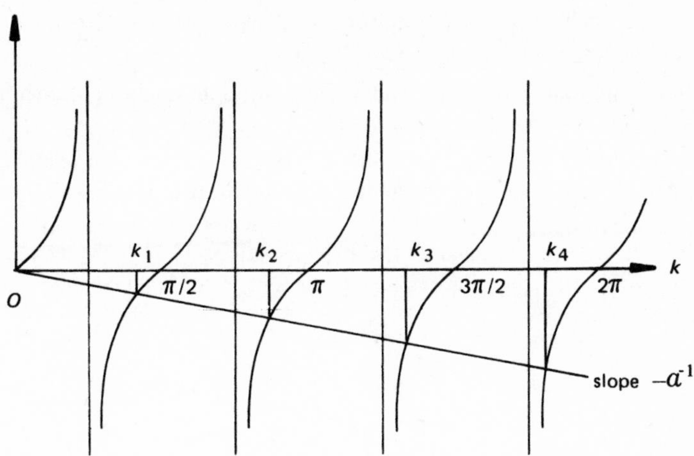

Fig. 4.1

We now form the superposition

$$u(x, t) = \sum_{n=1}^{\infty} \sin k_n x \, (C_n \cos k_n ct + D_n \sin k_n ct) \quad . \tag{4.12}$$

The initial conditions (4.9) then take the form

$$a(x) = \sum_{n=1}^{\infty} C_n \sin k_n x, \quad b(x) = \sum_{n=1}^{\infty} n D_n \sin k_n x \quad .$$

The theorem tells us that these expansions of $a(x)$ and $b(x)$ exist if $a(x)$ and $b(x)$ are piecewise continuous. The coefficients C_n are given by

$$\tfrac{1}{2}\left(L + \frac{\alpha}{\alpha^2 + k_n^2}\right) C_n = \int_0^L a(x) \sin k_n x \, dx \quad , \tag{4.13}$$

and a similar formula is obtained for D_n. Then (4.12) is the required solution.

4.2.3 Singular Sturm-Liouville problems

If $p(x)$ vanishes at either end of the interval, or if the interval is unbounded, the problem is said to be **singular**. At a singular boundary the condition on the solution $X(x)$ may be that $X(x)$ should be continuous, or be differentiable, or perhaps become infinite to less than a prescribed order. The classification of singular problems and their appropriate boundary conditions is beyond the scope of this book (for an indication of the approach to this problem see Coddington and Levinson (1955)). Here we confine ourselves to the case of a finite interval, with the following condition at any singular boundary:

At the boundary, $p(x) = 0$, and X and X' are continuous (and therefore bounded). (4.14)

With this restriction we now prove parts (b) and (c) of the fundamental theorem of section 4.2.1, for both the regular and the singular cases.

4.2.4 Properties of the eigenvalues and eigenfunctions

Let λ and μ be any two eigenvalues, possessing eigenfunctions $\phi(x)$ and $\psi(x)$ respectively. Then

$$\begin{aligned}(p\phi')' - q\phi + \lambda r\phi &= 0,\\ (p\psi')' - q\psi + \mu r\psi &= 0,\end{aligned} \tag{4.15}$$

and ϕ, ψ satisfy the boundary conditions (4.4) or (4.14) as appropriate.

We multiply the first of these equations by ψ and the second by ϕ, and subtract. We find that

$$(p\phi\psi' - p\psi\phi')' = (\lambda - \mu)\, r\phi\psi \quad . \tag{4.16}$$

From (4.4) or (4.14) we see that

$$p(\phi\psi' - \psi\phi') = 0 \text{ at both boundaries.} \tag{4.17}$$

It follows that the **eigenvalues are simple**. For, putting $\mu = \lambda$ and integrating (4.16), and using (4.17), we have $p(\phi\psi' - \psi\phi') = 0$. As $p(x) \not\equiv 0$ we deduce that $\phi\psi' = \psi\phi'$, whence $\phi = A\psi$, where A is constant. Hence λ cannot have two linearly independent eigenfunctions.

Now, integrating (4.16) from 0 to L and using (4.17), we obtain

$$(\lambda - \mu) \int_0^L r\phi\psi \, dx = 0 \quad . \tag{4.18}$$

We can now show that the **eigenvalues are real**. For, taking complex conjugates of (4.15) and (4.4) or (4.14) and noting which quantities are real, we see that $\bar{\phi}$ is an eigenfunction belonging to the eigenvalue $\bar{\lambda}$. Substituting $\bar{\lambda}$ for μ and $\bar{\phi}$ for ψ in (4.18) and observing that $r\phi\bar{\phi} = r|\phi|^2 \geqslant 0$ for $0 < x < L$, but is not identically zero, we see that the integral is positive. Hence $\lambda = \bar{\lambda}$, that is, λ is real.

It now follows that the **eigenfunctions are real**, except possibly for a complex constant factor. For, if the real eigenvalue λ possesses the eigenfunction $\phi(x) = \xi(x) + i\eta(x)$, where ξ and η are real, we may separate the real and imaginary parts of (4.15) and (4.4) or (4.14) and conclude that $\xi(x)$ and $\eta(x)$ are both eigenfunctions belonging to the simple eigenvalue λ. Hence $\eta(x) = A\xi(x)$ and we may write $\phi(x) = (1 + iA)\,\xi(x)$, where A is constant.

The **orthogonal property of the eigenfunctions** follows directly from (4.18). For, if $\mu \neq \lambda$ we have

$$\int_0^L r\phi\psi \, dx = 0 \quad .$$

Hence the functions $r^{1/2}\phi$ and $r^{1/2}\psi$ are orthogonal. Another way of describing this relation is to say that ϕ and ψ are **orthogonal with weight function** $r(x)$ in the interval $(0, L)$. Since all the eigenvalues are simple, any two distinct (that is, linearly independent) eigenfunctions are orthogonal with weight r.

4.3 SMALL TRANSVERSE VIBRATIONS OF A SUSPENDED STRING

Consider a string of length L and uniform line density ρ suspended from a fixed point. We take the x-axis to coincide with the vertical equilibrium position, with $x = 0$ at the lower end of the string and $x = L$ at the point of suspension. The equilibrium tension at a point P is then equal to the weight of the string below P, that is,

$$T(x) = \rho g x \quad .$$

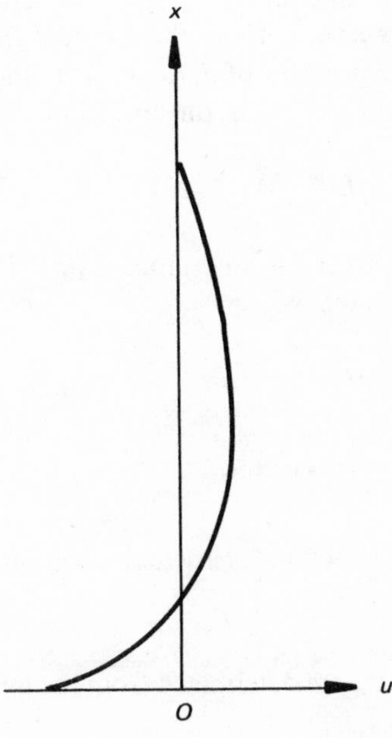

Fig. 4.2

Let the horizontal displacement of the string from the equilibrium position be $u(x, t)$, as shown in Fig. 4.2. Then the equation (4.1) of small transverse motion becomes

$$u_{tt} = g(xu_x)_x \quad . \tag{4.19}$$

This must be solved subject to $u(L, t) = 0$ and $u(0, t)$ bounded for all t. Seeking normal solutions of the form $X(x)\, e^{-i\omega t}$, we see that the normal modes of vibration are the solutions of the equation

$$g(x\,X')' + \omega^2 X = 0$$

which satisfy the boundary conditions $X(L) = 0$ and $X(0)$ is finite. This is a singular Sturm-Liouville problem. The eigenfunctions can be expressed in terms of **Bessel functions** by making the transformation $s = \sqrt{x}$.

Writing $\phi(s) = X(x)$ we obtain

$$\frac{d}{ds}\left[s\phi'(s)\right] + \lambda s\phi(s) = 0 \quad , \tag{4.20}$$

where $\lambda = 4\omega^2/g$, with $\phi(\sqrt{L}) = 0$ and $\phi(0)$ finite. Equations (4.20) is Bessel's equation of order zero, with solutions $J_0(s\sqrt{\lambda})$ and $Y_0(s\sqrt{\lambda})$, but we discard the latter to make $\phi(0)$ finite. Since $\phi(\sqrt{L}) = 0$, λ must satisfy $J_0(\lambda^{1/2} L^{1/2}) = 0$. Hence the eigenvalues are $\lambda_n = \mu_n^2 L^{-1}$, where the constants μ_n ($n = 1, 2, \ldots$) are the positive roots in ascending order of $J_0(\mu) = 0$, a known set of numbers. The corresponding eigenfunctions (that is, the normal modes) are given by

$$X_n(x) = J_0(\mu_n L^{-1/2} x^{1/2}) \quad .$$

Fig. 4.2 shows the form of the second mode $X_2(x)$. The frequency of vibration of the nth normal mode is ω_n, where

$$\omega_n = \tfrac{1}{2}\mu_n \sqrt{(g/L)} \quad .$$

The general solution for $u(x, t)$ is thus

$$u(x, t) = \sum_{n=1}^{\infty} J_0(\mu_n L^{-1/2} x^{1/2})\,(C_n \cos \omega_n t + D_n \sin \omega_n t) \quad , \tag{4.21}$$

where the constants C_n, D_n are determined from the initial conditions

$$u(x, 0) = a(x) \text{ and } u_t(x, 0) = b(x) \quad .$$

We set $t = 0$ in (4.21) to obtain

$$a(x) = \sum_{n=1}^{\infty} C_n J_0(\mu_n L^{-\frac{1}{2}} x^{\frac{1}{2}}) \quad .$$

By using the properties, and the orthogonality, of Bessel functions we easily deduce that

$$C_n = \frac{1}{L J_1^2(\mu_n)} \int_0^L a(x)\, J_0(\mu_n L^{-\frac{1}{2}} x^{\frac{1}{2}})\, dx \tag{4.22}$$

Differentiating (4.21) and setting $t = 0$, we similarly obtain

$$D_n = \frac{2}{\sqrt{(gL)}\mu_n J_1^2(\mu_n)} \int_0^L b(x)\, J_0(\mu_n\, L^{-\frac{1}{2}} x^{\frac{1}{2}})\, dx \tag{4.23}$$

4.3.1 Example

The linear density of suspended string of length L is $\rho(x) = \alpha x^m$, where α and m are constant and $m > 1$, and $x = 0$ at the free end. Show that the normal modes of vibration have the form $x^{-m/2}\, J_m(\gamma_n \sqrt{x})$, where γ_n is a root of $J_m(\gamma\sqrt{L}) = 0$. Find also the frequency of vibration of the nth normal mode.

SOLUTION

The equation of motion is $\rho u_{tt} = (T u_x)_x$

and the tension $T(x) = \displaystyle\int_0^x \rho(\xi)\, g d\xi = \frac{g \alpha x^{m+1}}{m+1}$.

Hence $$u_{xx} + \frac{m+1}{x} u_x = \frac{m+1}{gx} u_{tt} \quad .$$

For a normal mode $X(x)$ vibrating with frequency ω we have $u(x, t) = X(x)\, e^{-i\omega t}$ and so

$$X''(x) + \frac{1}{x}(m+1)\, X'(x) + \frac{(m+1)\,\omega^2}{gx} X(x) = 0 \quad .$$

Then if $X(x) = \phi(s)$, where $s = \sqrt{x}$, this equation reduces to

$$\phi''(s) + \frac{2m+1}{s} \phi'(s) + \gamma^2 \phi(s) = 0 \quad ,$$

where $\gamma^2 = 4(m+1)\, \omega^2 g^{-1}$.

The given solution suggests a further transformation $W(s) = s^m \phi(s)$. This gives

$$s^2 W''(s) + sW'(s) + (\gamma^2 s^2 - m^2)\, W(s) = 0 \quad ,$$

which is Bessel's equation of order m. Thus

$$W(s) = A\, J_m(\gamma s) + B\, Y_m(\gamma s) \quad .$$

The boundary conditions are: $u(0, t)$ is bounded and $u(L, t) = 0$, implying that $W(0)$ is finite and $W(\surd L) = 0$.

Hence $B = 0$ and $J_m(\gamma \surd L) = 0$. Consequently the normal modes have the form

$$X_n(x) = \frac{J_m(\gamma_n \surd x)}{x^{m/2}}$$

and γ_n is a root of $J_m(\gamma \surd L) = 0$. The frequency of vibration of the nth normal mode is ω_n, where

$$\omega_n = \frac{\gamma_n}{2} \sqrt{\left(\frac{g}{m+1}\right)} \quad .$$

4.3.2 Forced vibrations of a suspended string

Suppose a horizontal force $\rho f(x, t)$ per unit length is applied to a suspended uniform string. Then the equation of the resulting forced motion is

$$u_{tt} = g(xu_x)_x + f(x, t) \quad . \tag{4.24}$$

We wish to solve this equation subject to the boundary conditions

$$u(0, t) \text{ is bounded}, u(L, t) = 0 \tag{4.25}$$

and the initial conditions $\quad u(x, 0) = a(x), \quad u_t(x, 0) = b(x).$ (4.26)

We apply the method of section 3.4, that is, we seek an expansion in the normal modes of the system:

$$u(x, t) = \sum_{n=1}^{\infty} S_n(t)\, X_n(x) \ , \tag{4.27}$$

where $X_n(x) = J_0(\mu_n L^{-\frac{1}{2}} x^{\frac{1}{2}})$. We expand the three given functions f, a and b in the normal modes:

$$f(x, t) = \sum_{n=1}^{\infty} F_n(t)\, X_n(x), a(x) = \sum_{n=1}^{\infty} \alpha_n\, X_n(x), b(x) = \sum_{n=1}^{\infty} \beta_n\, X_n(x) \quad .$$

Substituting the expansions into (4.24), remembering that $g(x X_n')' + \omega_n^2 X_n = 0$, where $4L\omega_n^2 = g\mu_n^2$, and equating the coefficients of $X_n(x)$, we find

$$S_n''(t) + \omega_n^2 S_n(t) = F_n(t), n = 1, 2, \ldots \ .$$

These equations must be solved subject to the initial conditions

$$S_n(0) = \alpha_n, \ S_n'(0) = \beta_n \ .$$

The method of solution is chosen according to convenience, but a general formula can be obtained using the methods of section 1.3. The result is

$$S_n(t) = \frac{1}{\omega_n} \int_0^t F_n(\tau) \sin \omega_n(t - \tau)\, d\tau + \alpha_n \cos \omega_n t + \frac{\beta_n}{\omega_n} \sin \omega_n t \quad ,$$

in which $$F_n(\tau) = \frac{1}{L J_1^2(\mu_n)} \int_0^L f(x, \tau)\, J_0(\mu_n L^{-\frac{1}{2}} x^{\frac{1}{2}})\, dx$$

and α_n, β_n are given by the same formula with $a(x)$, $b(x)$ substituted for $f(x, \tau)$; $\omega_n = \frac{1}{2}\mu_n \sqrt{(g/L)}$ and the quantities μ_n are the positive zeros of $J_0(\mu)$. The solution of the problem is then

$$u(x, t) = \sum_{n=1}^{\infty} S_n(t)\, J_0(\mu_n L^{-\frac{1}{2}} x^{\frac{1}{2}}) \quad .$$

4.3.3 Small transverse vibrations of a rotating string

Let us now examine the vibrations of a uniform string, of line density ρ and length L, which is fixed at one end O and is rotating freely around O. When gravity and air resistance are neglected, the quasi-equilibrium configuration of the string is a straight line rotating with constant angular speed Ω in a plane passing through O. The string may vibrate around this equilibrium configuration, and we denote the transverse displacement of the point distant x from O at time t by $u(x, t)$.

The tension $T(x)$ at the point distant x from O is equal to the sum of the forces acting on all elements of the string from the point x to its free end $x = L$. So

$$T(x) = \int_x^L \rho\, \Omega^2 \xi d\xi = \tfrac{1}{2}\rho\, \Omega^2 (L^2 - x^2) \quad ,$$

and the free vibrations of the rotating string are governed by the equation

$$u_{tt} = \tfrac{1}{2}\Omega^2 [(L^2 - x^2)\, u_x]_x \quad . \tag{4.28}$$

The boundary conditions are

$$u(0, t) = 0, \; u(L, t) \text{ bounded.} \tag{4.29}$$

By seeking normal solutions of the form $u(x, t) = X(x)\, e^{-i\omega t}$ we see that the normal modes of vibration are the solutions of

$$\Omega^2 [(L^2 - x^2)\, X'(x)]' + 2\omega^2 X(x) = 0 \tag{4.30}$$

which satisfy the conditions

$$X(0) = 0 \text{ and } X(L) \text{ finite.} \tag{4.31}$$

Setting $x = Ls$ and $X(x) = \phi(s)$ we obtain

$$[(1 - s^2)\phi']' + \lambda\phi = 0 \quad , \tag{4.32}$$

where $\lambda = 2\omega^2/\Omega^2$, with

$$\phi(0) = 0 \text{ and } \phi(1) \text{ finite} \quad . \tag{4.33}$$

Equation (4.32) is **Legendre's equation**. The only solutions which satisfy (4.33) are the **odd Legendre polynomials** $P_{2n-1}(s)$ $(n = 1, 2, \ldots)$ (see section 13.3.1). Hence the eigenvalues are $\lambda_n = (2n-1)2n$, the normal frequencies are given by $\omega_n^2 = n(2n-1)\Omega^2$, and the corresponding normal modes are $X_n(x) = P_{2n-1}(x/L)$.

Consequently the general displacement of the rotating string is

$$u(x, t) = \sum_{n=1}^{\infty} P_{2n-1}(x/L)\; [C_n \cos\sqrt{(2n^2 - n)}\, \Omega t + D_n \sin\sqrt{(2n^2 - n)}\Omega t] \tag{4.34}$$

where the constants C_n and D_n are determined from the initial conditions $u(x, 0) = a(x)$, $u_t(x, 0) = b(x)$. We require the expansions

$$a(x) = \sum_{n=1}^{\infty} C_n P_{2n-1}(x/L) \text{ and } b(x) = \sum_{n=1}^{\infty} D_n\sqrt{(2n^2 - n)}\Omega\, P_{2n-1}(x/L)$$

From the properties of the Legendre polynomials we deduce that

$$C_n = \frac{4n-1}{L} \int_0^L a(x)\, P_{2n-1}(x/L)\, dx$$

and

$$D_n = \frac{4n-1}{L\Omega\sqrt{(2n^2 - n)}} \int_0^L b(x) P_{2n-1}(x/L)\, dx \quad .$$

The required solution is then obtained by substituting these expressions into (4.34).

Example
Suppose the initial conditions are

$$u(x, 0) = \alpha x^3 + \beta x, \ \alpha, \beta \text{ constant},$$

and $$u_t(x, 0) = 0 \quad .$$

As $u_t(x, 0) = 0$, we have immediately that $D_n = 0$ for all n.
So

$$u(x, t) = \sum_{n=1}^{\infty} C_n P_{2n-1}(x/L) \cos\sqrt{(2n^2 - n)}\, \Omega t.$$

Now

$$\alpha x^3 + \beta x = \tfrac{2}{5} L^3 \alpha P_3(x/L) + (\tfrac{3}{5} L^3 \alpha + L\beta) P_1(x/L) \quad ,$$

and so the first initial condition gives

$$(\tfrac{3}{5}\alpha L^3 + \beta L) P_1(x/L) + \tfrac{2}{5}\alpha L^3 P_3(x/L) = \sum_{n=1}^{\infty} C_n P_{2n-1}(x/L) \quad .$$

Equating coefficients, we obtain

$$C_1 = (\tfrac{3}{5}\alpha L^3 + \beta L),\ C_2 = \tfrac{2}{5}\alpha L^3,\ C_n = 0 \text{ when } n > 2 \quad .$$

Hence

$$u(x, t) = (\tfrac{3}{5}\alpha L^3 + \beta L) P_1(x/L) \cos \Omega t + \tfrac{2}{5}\alpha L^3 P_3(x/L) \cos\sqrt{6}\,\Omega t \quad .$$

PROBLEMS

4.1 Not every Sturm-Liouville problem leads to simple eigenvalues. Consider the problem of **periodic boundary conditions**:

$$\phi'' + \lambda\phi = 0,\ 0 < x < 2\pi$$

with

$$\phi(0) = \phi(2\pi) \text{ and } \phi'(0) = \phi'(2\pi)$$

Show that the eigenvalues are $\lambda_0 = 0$ (simple) and $\lambda_n = n^2 (n = 1, 2, \ldots)$ (all double). Find the corresponding real eigenfunctions (that is, one for λ_0 and two linearly independent eigenfunctions for each λ_n), and show that they can be chosen to be mutually orthogonal. Observe that here a modification of the theorem of section 4.2.1 corresponds to the theorem on Fourier series of section 3.2.2.

4.2 Two uniform strings AO and OB with line densities ρ_1 and ρ_2 are joined at O. The ends A and B are fixed so that $AB = 2L$ and $OA = OB = L$, and the tension is T. A particle of mass M is attached at O. Show that the frequencies ω of the normal transverse vibrations are the solutions of the equation

$$\frac{1}{c_1}\cot\frac{\omega L}{c_1} + \frac{1}{c_2}\cot\frac{\omega L}{c_2} = \frac{M\omega}{T} ,$$

where $c_1^2 = T/\rho_1$ and $c_2^2 = T/\rho_2$.

4.3 Show that the normal modes of the problem $c^2 u_{xx} = u_{tt}$, $u_x(0,t) - \alpha u(0,t) = 0, u_x(L, t) + \beta u(L, t) = 0, \alpha > 0, \beta > 0$, may be written

$$X_n(x) = \sqrt{2}(k_n^2 + \alpha^2)^{-\frac{1}{2}}(k_n \cos k_n x + \alpha \sin k_n x) ,$$

where $$(k_n^2 - \alpha\beta) \tan k_n L = (\alpha + \beta) k_n .$$

Show that $$\int_0^L X_n^2(x)\,\mathrm{d}x = M_n \text{ where } M_n = L + \frac{(\alpha + \beta)(k_n^2 + \alpha\beta)}{(k_n^2 + \alpha^2)(k_n^2 + \beta^2)} ,$$

and hence show that, for $0 < x < L$,

$$1 = 2\sum_{n=1}^{\infty} \frac{k_n \sin k_n L + \alpha(1 - \cos k_n L)}{M_n k_n (k_n^2 + \alpha^2)} (k_n \cos k_n x + \alpha \sin k_n x) .$$

4.4 A string of length L is suspended from the end $x = L$ and a particle of mass $2m$ is fastened to the other end $x = 0$. The linear density of the string is $\rho(x) = m(L^2 + Lx)^{-\frac{1}{2}}$.

Show that the equation of small transverse motion of the string about its vertical equilibrium position is

$$\frac{\partial^2 u}{\partial s^2} = \frac{2}{g}\frac{\partial^2 u}{\partial t^2} , \quad \text{where } s = \sqrt{(2L)} - \sqrt{(L + x)}.$$

4.5 A uniform string of length L and line density ρ is suspended from the end $x = L$, and a particle of mass ρh, where h is a constant, is fastened to the end $x = 0$. Find the equation satisfied by the normal frequencies of transverse vibration of the string.

4.6 A string of length L and tension T has its ends fixed at the points $x = 0$ and $x = L$. Given that the line density of the string is $Tc^{-2}(1 + kx)^{-2}$, where c and k are constants, show that the frequencies ω_n of the transverse normal vibrations are given by

$$\frac{\omega_n{}^2}{k^2c^2} = \frac{1}{4} + \left\{\frac{\pi n}{\ln(1+kL)}\right\}^2, \quad n = 1, 2, \ldots .$$

4.7 A string with line density $m(2L^2 - x^2)^{-1/2}$, where m and L are constants, is fastened at the end $x = 0$ to a fixed point A and has a particle of mass m attached to the other end $x = L$. Show that when the string rotates about A at a constant angular speed Ω, the equation for its small transverse vibrations is

$$\frac{\partial^2 u}{\partial t^2} = \Omega^2 \frac{\partial^2 u}{\partial s^2} \quad ,$$

where $s = \arcsin(x/L\sqrt{2})$.

4.8 A pliable thread of length L and line density kx^2 hangs vertically from its end $x = L$ which is fixed. Show that the equation for small transverse vibrations of the thread about its vertical equilibrium position is

$$\frac{\partial^2 u}{\partial x^2} + \frac{3}{x}\frac{\partial u}{\partial x} = \frac{3}{gx}\frac{\partial^2 u}{\partial t^2} \quad .$$

Deduce that the normal modes are

$$x^{-1} J_2(\mu_n\sqrt{(x/L)}), n = 1, 2, \ldots \quad ,$$

where μ_n is a positive root of the equation $J_2(\mu) = 0$. Find also the frequencies of the normal vibrations.

4.9 A string of length $2l$ is fixed at its ends and stretched at a tension kp^2. The density of the string at a displacement x from its mid-point O is

$$\rho(x) = \frac{k}{(2l - |x|)^2} \quad .$$

If the string can execute normal transverse oscillations of the form

$$u(x, t) = U(x) \cos \omega t \quad ,$$

where

$$U(x) = U_1(x), \quad x > 0$$
$$= U_2(x), \quad x < 0 \quad ,$$

find a differential equation for $U_1(x)$.

Verify that $U_1(x) = A\left(2 - \frac{x}{l}\right)^{\frac{1}{2}} \sin\left\{\lambda \ln\left(2 - \frac{x}{l}\right)\right\}$ satisfies the equation where $\lambda^2 = \frac{\omega^2}{p^2} - \frac{1}{4}$ and A is constant, and that $U_1(l) = 0$.

Obtain by considerations of symmetry an expression for $U_2(x)$, and deduce that the frequencies ω of the normal modes are given by the solutions of the equation

$$\tan(\lambda \ln 2) + 2\lambda = 0$$

together with the values

$$p\left\{\frac{1}{4} + \left(\frac{n\pi}{\ln 2}\right)^2\right\}^{\frac{1}{2}}, \; n = 1, 2, \ldots .$$

4.10 A uniform string OA of length L and density ρ hangs vertically from a small smooth light ring A which is free to slide on a fixed horizontal rod. The ring A is subjected to a horizontal force $F\cos\omega t$, and the string executes small transverse motion in which its displacement is $u(x, t)$. Show that $T_A u_x = F\cos\omega t$, where T_A is the tension at A. Deduce that the resulting *forced motion* (in the sense of section 3.5) is

$$u(x, t) = -\frac{F J_0[2\omega\surd(x/g)] \cos\omega t}{\rho\omega\surd(gL) J_1[2\omega\surd(L/g)]} \quad .$$

CHAPTER 5

Dispersion

5.1 DISPERSIVE MEDIUM

It has been shown in Chapter 3 that any solution (subject to certain conditions) of the one-dimensional classical wave equation may be expressed as a linear combination of the harmonic wave solutions by means of Fourier analysis. The statement

(i) that a medium transmits all linear combinations of harmonic waves each travelling at a fixed speed c

is therefore almost equivalent to the statement

(ii) that the medium is governed by the equation $c^2 u_{xx} = u_{tt}$.

Statement (i) does not include all the solutions of the equation, for instance it omits the solution $u = xt$, but it does, as we have seen, provide those needed in most applications.

A dispersive medium is a system which transmits harmonic waves of differing frequency at different speeds. Such media clearly invite the use of a suitable modification of statement (i) for their description. We study wave propagation in terms of known harmonic wave solutions rather than directly from the governing equations which, though simple for some dispersive media, are complicated for others, notably surface waves on water and electromagnetic waves in a dielectric.

DEFINITION

A one-dimensional linear dispersive medium is a system which transmits all linear combinations of harmonic waves of the form

$$\cos(kx - \omega t - \epsilon), \tag{5.1}$$

where ω and k are real numbers which satisfy a given relation

$$D(\omega, k) = 0 \quad , \tag{5.2}$$

called the **dispersion relation** of the medium.

The medium thus described is also uniform in space and constant in time. The definition can be extended to non-uniform or changing media by allowing D to depend on x or t. It can also be extended to non-linear systems.

5.2 GROUP VELOCITY

In a dispersive medium the wave pattern continually changes. Consider the linear superposition of two harmonic waves

$$u(x, t) = \cos(k_1x - \omega_1t) + \cos(k_2x - \omega_2t) \quad ,$$

in which k_1, k_2, ω_1, ω_2 are all positive.
When written in the form

$$u(x, t) = 2\cos\left(\frac{k_1+k_2}{2}x - \frac{\omega_1+\omega_2}{2}t\right)\cos\left(\frac{k_2-k_1}{2}x - \frac{\omega_2-\omega_1}{2}t\right)$$

it is seen to be equivalent to a harmonic wave travelling with velocity $\dfrac{\omega_1+\omega_2}{k_1+k_2}$ modulated by another harmonic wave with velocity $\dfrac{\omega_2-\omega_1}{k_2-k_1}$. At any instant the wave form is as shown in Fig. 5.1. The variation in amplitude depends on the modulating wave, whose outline is shown touching all the individual peaks.

Let $\omega(k)$ denote the **dispersion function**, which satisfies the dispersion relation $D[\omega(k), k] = 0$.

Fig. 5.1

For $|k_2 - k_1| \ll k_1$ the velocities of the wave and its modulation can be approximated by $V(k_1)$ and $U(k_1)$, where

$$V(k) = \frac{\omega(k)}{k}$$, called the **phase velocity**

and

$$U(k) = \frac{d\omega}{dk}$$, called the **group velocity** .

If dispersion is absent, $V = U$, and the whole pattern moves rigidly. In a dispersive medium $V \neq U$, and the individual wave crests move with the phase velocity V, varying in height as they pass through the modulation outline which is itself moving with the group velocity U. Such behaviour is not confined to periodically modulated waves, but occurs in any bundle of waves of nearly equal wavelengths. It is observable, for example, in a group of surface waves on deep water, where $U = \frac{1}{2}V$ approximately. Each individual wave is seen to arise at the rear and advance through the group, moving at twice the speed of the group as a whole and gradually diminishing as it approaches the front.

The group velocity is physically more significant than the phase velocity. It is the velocity of transmission of slowly varying modulations of a harmonic wave, and indeed in many applications it can be regarded as the effective signal velocity. In most physical systems energy is transmitted at the group velocity.

5.3 EXAMPLE: THE ELASTICALLY ANCHORED STRING

5.3.1 The equation of motion

A uniform stretched string of density ρ is anchored elastically to its equilibrium position by a transverse restoring force per unit length proportional to the displacement (Fig. 5.2). The equation of small transverse motion is

$$c^2 u_{xx} - \beta^2 u = u_{tt} \quad , \tag{5.3}$$

where ρc^2 is the tension and $\rho\beta^2$ is the restoring force constant. (This equation also occurs in quantum mechanics; it is called the Klein-Gordon equation.)

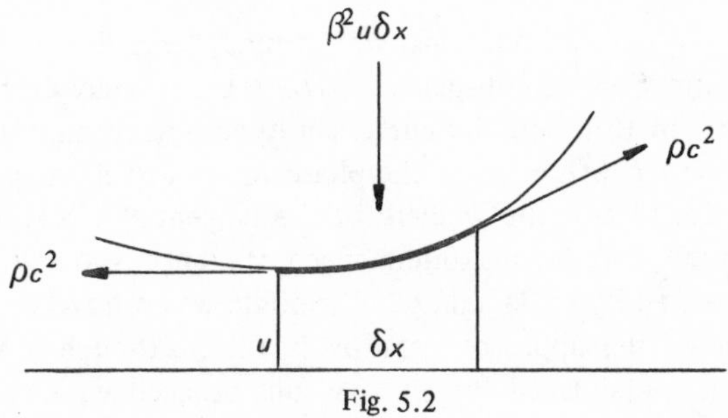

Fig. 5.2

5.3.2 The dispersion relation

Because the equation is linear and has constant coefficients the string constitutes a uniform constant linear medium. It is convenient to consider a harmonic wave of the form

$$\psi(x, t) = Ae^{i(kx - \omega t)} \quad ,$$

where A is a complex constant, which is a linear combination of two waves such as (5.1) and therefore admissible in a linear medium. The wave ψ is a solution provided that

$$D(\omega, k) \equiv \omega^2 - c^2k^2 - \beta^2 = 0 \quad . \tag{5.4}$$

This is the dispersion relation for the string (Fig. 5.3).
The phase velocity $V(k) = \omega/k = c(1 + \beta^2 c^{-2} k^{-2})^{1/2}$.
The group velocity $U(k) = d\omega/dk = c(1 + \beta^2 c^{-2} k^{-2})^{-1/2}$.

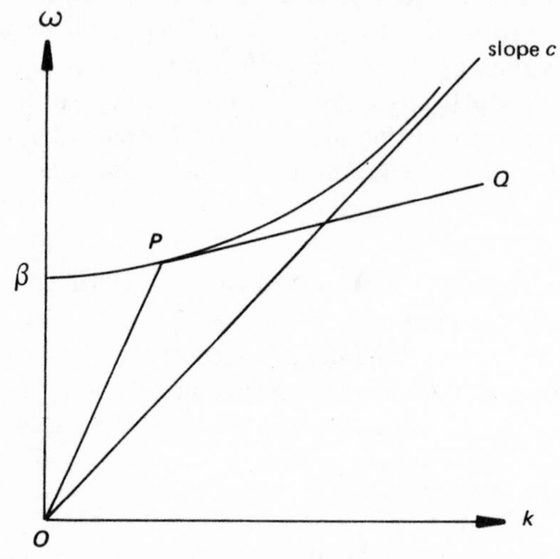

Fig. 5.3

When discussing dispersion relations we normally regard ω as positive and $\omega(k)$ as an even function. The dispersion curve is accordingly drawn only in the positive quadrant. In this case the curve is a hyperbola asymptotic to the line $\omega = ck$. At a point P of the curve the phase velocity V is the gradient of OP and the group velocity U is the gradient of the tangent PQ. Now in Chapter 11 it will be shown that c is the maximum speed of transmission of a disturbance in a medium governed by (5.3), and yet harmonic waves travel at speeds $V > c$. The explanation of the apparent paradox is that, although a wave such as $\cos k(x - Vt)$, once established for all x, would proceed with velocity V, the velocity of the front F of a finite wave train (Fig. 5.4) is limited to c. The main

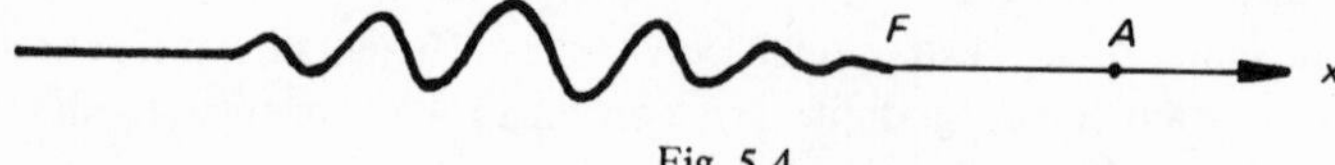

Fig. 5.4

body of the wave train, which can be regarded as a signal, travels with the group velocity $U(< c)$ so that, as the signal proceeds, the front outpaces it. At a fixed point A low-amplitude precursors will be observed before the main signal arrives.

Waves of any length can propagate, but their frequency must be at least β. The string is therefore a **high-pass system**, with a **cut-off frequency** β. Writing

$k = ip$ we obtain $c^2p^2 = \beta^2 - \omega^2$. Hence, for $\omega < \beta$, there are solutions of the form $e^{-px}e^{-i\omega t}$, where p is real. The profile is an exponential curve, all points move in unison, and there are no travelling waves. No signals can be transmitted below the cut-off frequency, and $U \to 0$ as $\omega \to \beta$.

5.3.3 Energy transport

We now consider the transport of energy in a harmonic wave. The energy density in the system is

$$\mathcal{E}(x, t) = \tfrac{1}{2}\rho u_t^2 + \tfrac{1}{2}\rho c^2 u_x^2 + \tfrac{1}{2}\rho\beta^2 u^2 \quad .$$

The energy flux, defined as the rate of working of the force exerted at P by the tension in the portion of the string to the left of P, is

$$\mathcal{F}(x, t) = -\rho c^2 u_x u_t \quad .$$

It follows, with the help of (5.3), that

$$\frac{\partial \mathcal{E}}{\partial t} + \frac{\partial \mathcal{F}}{\partial x} = 0 \quad , \tag{5.5}$$

which is an expression of **conservation of energy** (which necessarily holds in this case because the force system is conservative).

For the wave $u = a\cos\theta$, where $\theta = kx - \omega t$,

$$\mathcal{E} = (\tfrac{1}{2}\rho\omega^2 + \tfrac{1}{2}\rho c^2 k^2)\, a^2 \sin^2\theta + \tfrac{1}{2}\rho\beta^2 a^2 \cos^2\theta \quad ,$$
$$\mathcal{F} = \rho c^2 k\omega a^2 \sin^2\theta \quad .$$

By integrating with respect to t over a period $2\pi/\omega$ of the oscillation, and using the dispersion relation (5.4), we obtain

the average energy density $\bar{\mathcal{E}} = \tfrac{1}{2}\rho\omega^2 a^2$
and the average flux $\bar{\mathcal{F}} = \tfrac{1}{2}\rho c^2 k\omega a^2$ (5.6)

The mean velocity of transport of energy is therefore

$$\bar{\mathcal{F}}/\bar{\mathcal{E}} = c^2 k/\omega = U \quad . \tag{5.7}$$

Hence, in the elastically anchored string, energy is transported at the group velocity. It is of interest to note that the quantity $d\omega/dk$ has physical significance even in the case of a harmonic wave, in which k is fixed.

5.4 MATURE DISPERSION

5.4.1 Wave trains

When a stone falls into a lake, a complicated motion spreads outwards, but after some time the waves observed are almost sinusoidal. The various harmonic components of the initial disturbance, travelling at different speeds, have become dispersed so that waves of different frequency are eventually found in different parts of the lake. In this section we consider a one-dimensional medium, and treat the case when the dispersion process has developed to a stage at which the wave pattern can be closely approximated over an extensive region by a **wave train**, that is, a modified harmonic wave with slowly varying amplitude, frequency and wavelength. We may represent the wave train by $u = \text{Re}\psi$, where

$$\psi(x, t) = a(x, t)e^{i\theta(x, t)} \quad , \tag{5.8}$$

where $a(x, t)$ is the amplitude and $\theta(x, t)$ is the phase.

The wave train is neither an exact solution of the equations governing the motion nor a linear combination of harmonic waves satisfying the dispersion relation. It is an asymptotic approximation to an exact solution, and the development of a linear combination of harmonic waves into a form approximated by a wave train is studied in section 5.7.3.

5.4.2 Local wave number and frequency

We wish to define, in the neighbourhood of any point (x_0, t_0), a local wave number k_0 and frequency ω_0 with the property that the phase is locally given by

$$\theta(x, t) \cong \theta(x_0, t_0) + k_0(x - x_0) - \omega_0(t - t_0) \quad .$$

This is achieved by defining the functions $k(x, t)$ and $\omega(x, t)$ as follows.

Wave number $\quad k(x, t) = \partial\theta/\partial x \qquad (5.9)$

and frequency $\omega(x, t) = -\,\partial\theta/\partial t. \qquad (5.10)$

Then the required quantities are $k_0 = k(x_0, t_0)$ and $\omega_0 = \omega(x_0, t_0)$.

We note that $\dfrac{\partial k}{\partial t} + \dfrac{\partial \omega}{\partial x} = 0, \qquad (5.11)$

which can be interpreted as the equation of conservation of waves, for the rate at which waves pass a fixed point, the wave flux, is $\omega/2\pi$, and $k/2\pi$ is the wave density; the velocity of transport of the waves is thus $\omega/k = V$ (cf. equations (5.5) and (5.7)).

5.4.3 Group lines and phase lines

Equation (5.11) may also be written

$$\frac{\partial k}{\partial t} + U(k)\,\frac{\partial k}{\partial x} = 0 \quad , \tag{5.12}$$

where $U(k) = \omega'(k)$, the group velocity.

This is a partial differential equation for the function $k(x, t)$, possessing characteristics which are solutions of the equation $\mathrm{d}x/\mathrm{d}t = U[k(x,t)]$. On each characteristic, (5.12) becomes $\mathrm{d}k/\mathrm{d}t = 0$, and so k is constant. Hence U also is constant and therefore the characteristics are straight lines, called **group lines** because the gradient of each line with respect to the t-axis is the value of the group velocity U on the line.

Given that $k(x_0, t_0) = k_0$, then $k = k_0$ at all points on the group line

$$x = x_0 + U(k_0)\,(t - t_0) \quad .$$

If $k(x, t)$ is known on an arc Γ in xt-space (Fig. 5.5), group lines can be constructed through all points of Γ and (5.12) can be solved in the region thus generated, bounded by the curve Γ and the group lines labelled k_1 and k_4 in the diagram. The **phase lines** (curves of constant θ) are the solution curves of the differential equation

$$\frac{\mathrm{d}x}{\mathrm{d}t} = \frac{\omega[k(x, t)]}{k(x, t)} \tag{5.13}$$

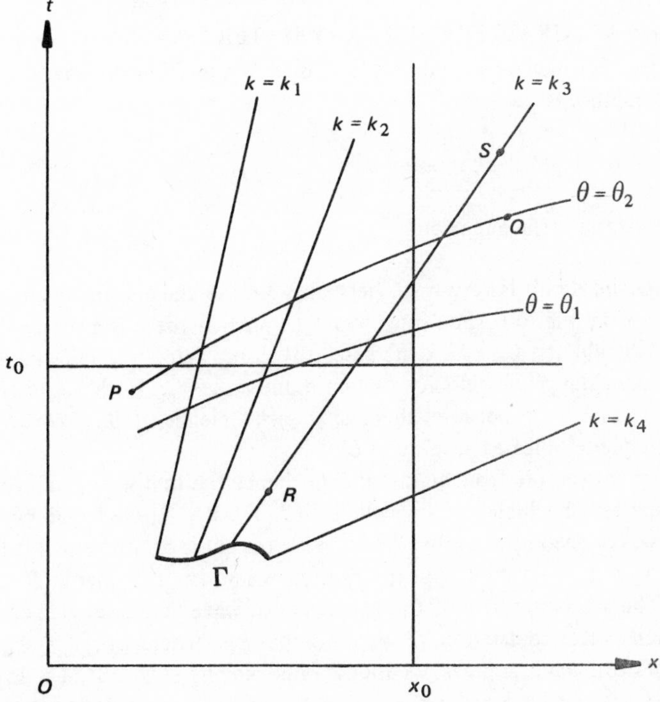

Fig. 5.5

The gradient of each curve is the phase velocity $V(k)$. Fig. 5.5 illustrates the case of gravity waves on deep water, for which the dispersion relation is $\omega^2 = gk$, giving $k = \frac{1}{4}gU^{-2}$, $\omega = \frac{1}{2}gU^{-1}$, $V = 2U$.

We see that k and ω decrease as U increases, so that $k_1 > k_2 > k_3 > k_4$ in the diagram, and the following conclusions can be drawn.

(i) At an instant t_0 the wavelength in the wave train increases with x.
(ii) The frequency of the waves reaching a fixed point x_0 increases with t.
(iii) An individual wave crest moves along a phase line such as PQ. It travels with steadily increasing velocity $V(k)$, and the wavelength steadily increases.
(iv) An observer following a group line, such as RS, moves with constant velocity $U(k_3)$, and sees waves of constant wavelength which overtake him at velocity $2U(k_3)$.

The method of this section is valid only when the dispersion is fully developed, that is, in regions where group lines belonging to neighbouring values of k do not meet. The treatment can be extended to non-uniform media, where the group lines become curved, and it can be applied to non-linear waves. It is a useful approach to some of the difficult problems connected with water waves (Whitham 1974).

5.5 EXAMPLE: GRAVITY-CAPILLARY WATER WAVES

In Chapter 10 it will be shown that the dispersion relation for waves on water of constant depth h is

$$\omega^2 = g(k + \tau k^3) \tanh kh \quad ,$$

where $\tau\rho g$ is the surface tension.

Unless the depth is very small (less than 5 mm), the dispersion curve has the form shown in Fig. 5.6. For long waves (k small), there is a maximum phase (and group) velocity V_M. As k increases the group velocity attains its minimum value U_0, and the phase velocity its minimum value V_m. Both V and U increase indefinitely as $k \to \infty$, because, then, $\omega^2 \sim g\tau k^3$. Hence, if $U_0 < U < V_M$, there are two frequencies for each value of U.

At a point remote from an initial disturbance the first waves to arrive would be fast ripples for which $k > k_A$, where $U(k_A) = V_M$. These would be followed by long waves accompanied by ripples with the same group velocity; the wave numbers k_B and k_E would appear together, where the tangents at B and E are parallel. The wave numbers of the two types of wave then approach each other, and the last waves to arrive have wave number k_D, where $U(k_D) = U_0$. Finally an area of calm water appears, its boundary advancing at velocity U_0. In practice the ripples are not observed preceding or accompanying the long waves because they are rapidly attenuated by viscous forces.

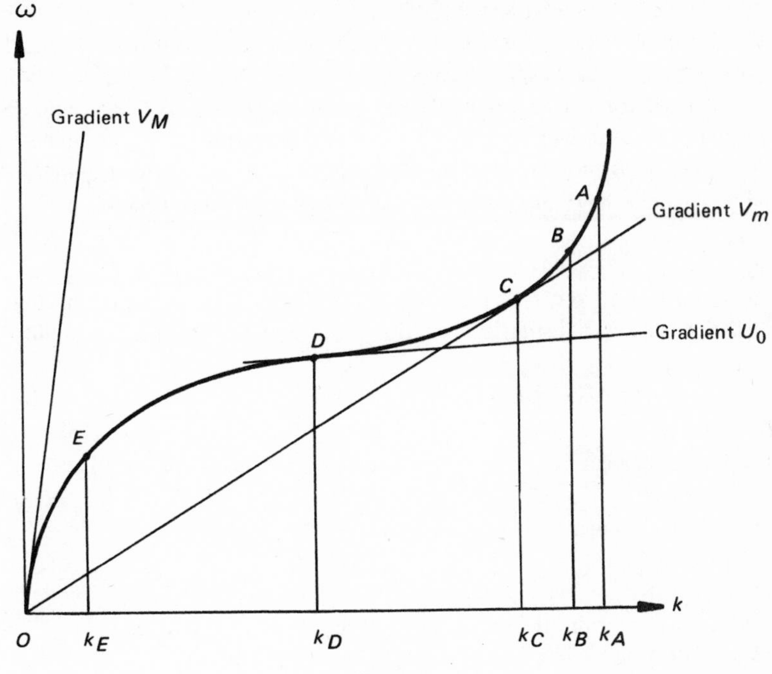

Fig. 5.6

We note that $U > V$ for ripples with $k > k_C$, so that a group of such ripples moves faster than the wave crests. This gives rise to the effect observed when there is a fixed obstacle in a steady stream flowing with speed V(Fig. 5.7).

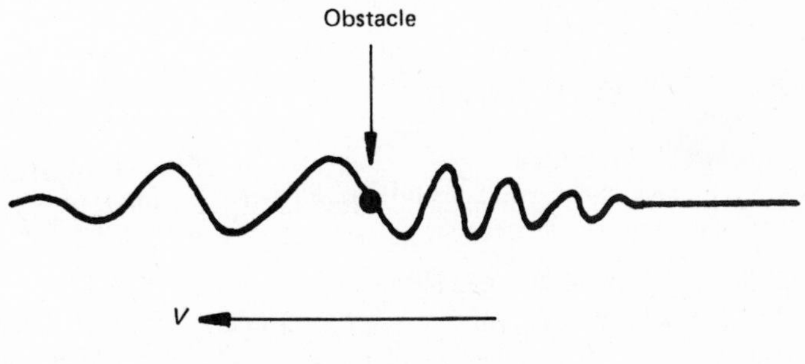

Fig. 5.7

The disturbance produces waves travelling at various speeds, but there will be appreciable amplitude only in those with phase velocity V upstream, which appear as static waves. If $V < V_\mathrm{m}$ no waves are seen. When $V_\mathrm{m} < V < V_\mathrm{M}$ there are two values of k for which V is the phase velocity. The ripples, for which $U > V$, appear upstream of the obstacle and the gravity waves ($U < V$) downstream.

5.6 BOUNDARY VALUE PROBLEMS

The discussion in this chapter has been mostly restricted, so far, to the behaviour of individual harmonic waves in a dispersive medium. A linear dispersive medium, as defined in section 5.1, can transmit any linear combination of harmonic waves satisfying the dispersion relation. We now turn our attention to such combinations.

In this section we show how boundary value problems in the dispersive case can be treated by direct application of the Fourier method of Chapter 3.

Example

It is required to find $u(x, t)$, a wave in a linear dispersive medium, occupying the region $0 \leqslant x \leqslant L$, with dispersion function $\omega(k)$, satisfying boundary conditions

$$u(0, t) = 0 = u(L, t) \tag{5.14}$$

and initial conditions

$$u(x, 0) = a(x), u_t(x, 0) = b(x) \quad . \tag{5.15}$$

Solution

We seek $u(x, t)$ as a linear combination of functions of the form $e^{i[kx-\omega(k)t]}$, k real.

Inspection of the boundary conditions shows that the required combination has the form

$$u(x, t) = \sum_{n=1}^{\infty} \sin \frac{n\pi x}{L} (A_n \cos\omega_n t + B_n \sin\omega_n t) \quad , \tag{5.16}$$

where $\omega_n = \omega(n\pi/L)$.

The initial conditions (5.15) then give

$$A_n = \frac{2}{L} \int_0^L a(x) \sin \frac{n\pi x}{L} \, dx \quad \text{and} \quad \omega_n B_n = \frac{2}{L} \int_0^L b(x) \sin \frac{n\pi x}{L} \, dx \quad .$$

The following observations can be made.

(i) Some of the solutions ω_n of the dispersion relation may be complex. The corresponding Fourier components will be damped oscillations or exponential decays.

(ii) By assuming k to be real we have found the unique *stationary harmonic wave combination* satisfying the conditions, but there may be other solutions. For example, the equation of flexural waves in a thin beam is of the fourth order in x:

$$u_{tt} + \gamma^2 u_{xxxx} = 0 \quad , \tag{5.17}$$

and therefore four boundary conditions are needed for the solution to be unique. Our solution (5.16) satisfies the extra conditions

$$u_{xx}(0, t) = 0 = u_{xx}(L, t) \quad .$$

Other sets of boundary conditions lead to the inclusion of imaginary values of k, so that $\cosh kx$ and $\sinh kx$ appear in the normal modes, and the assumption (5.16) must be duly modified.

5.7 THE INITIAL VALUE PROBLEM

5.7.1 Solution by the Fourier method

Let the functions $a(x)$ and $b(x)$, defined for all real values of x, satisfy the conditions of Fourier's Integral Theorem.

Problem

It is required to find, in a linear dispersive medium with dispersion function $\omega(k)$, a wave $u(x, t)$, $t \geqslant 0, x \in \mathbb{R}$, satisfying the initial conditions

$$u(x, 0) = a(x), u_t(x, 0) = b(x) \quad . \tag{5.18}$$

Solution

We suppose that u can be expressed as a linear combination of harmonic waves

$$u(x, t) = \tfrac{1}{2} \int_{-\infty}^{\infty} [A(\kappa)e^{i(\kappa x - \omega t)} + B(\kappa)e^{i(\kappa x + \omega t)}] \, d\kappa \quad , \tag{5.19}$$

where $A(\kappa)$ and $B(\kappa)$ take complex values.

The initial conditions give

$$a(x) = \tfrac{1}{2} \int_{-\infty}^{\infty} [A(\kappa) + B(\kappa)]e^{i\kappa x} d\kappa \quad ,$$

$$b(x) = \tfrac{1}{2} \int_{-\infty}^{\infty} i\omega(\kappa) \, [-A(\kappa) + B(\kappa)]e^{i\kappa x} d\kappa \quad .$$

Using the Fourier integral formula (3.52) we obtain

$$A(\kappa) + B(\kappa) = \frac{1}{\pi} \int_{-\infty}^{\infty} a(x) \, e^{-i\kappa x} \, dx \quad ,$$

$$A(\kappa) - B(\kappa) = \frac{i}{\pi\omega(\kappa)} \int_{-\infty}^{\infty} b(x) \, e^{-i\kappa x} \, dx \quad ,$$

whence

$$\left.\begin{array}{l}A(\kappa)\\B(\kappa)\end{array}\right\} = \frac{1}{2\pi}\int_{-\infty}^{\infty}\left\{a(x) \pm \frac{i}{\omega(\kappa)}\,b(x)\right\}e^{-i\kappa x}\,dx \quad . \tag{5.20}$$

The required solution is obtained by substituting the expressions for $A(\kappa)$ and $B(\kappa)$ into (5.19).

Adopting the convention that $\omega(-\kappa) = \omega(\kappa) \geqslant 0$ we note that, since $a(x)$ and $b(x)$ are real, $B(\kappa) = \bar{A}(-\kappa)$, so that (5.19) simplifies to $u(x, t) = \text{Re}\,\psi$,

where $$\psi(x, t) = \int_{-\infty}^{\infty} A(\kappa)\,e^{i[\kappa x - \omega(\kappa)t]}\,d\kappa \quad . \tag{5.21}$$

If there is an underlying partial differential equation governing the system, it may possess other solutions satisfying (5.18) which are not of the form (5.19) [cf. section 5.6, observation (ii)].

5.7.2 Wave packets

Consider first the case when $|A(\kappa)|$ is small compared with its mean value except for values of κ satisfying

$$|\kappa - k| < \Delta k \ll k \quad ,$$

where k is a fixed wave number (here assumed positive for convenience). Then (5.21) is called a **wave packet** centred on k, and $2\Delta k$ may be termed the **width** of the **spectrum** of the packet (Fig. 5.8a).

Initially $\psi(x, 0) = e^{ikx}\displaystyle\int_{-\infty}^{\infty} A(\kappa)e^{i(\kappa-k)x}d\kappa$. Fig. 5.8b shows the form of $u(x, 0)$. It can be regarded as a wave with wave number k modulated by the slowly varying function given by the integral, which is represented by the envelope of the wave.

Expanding $\omega(\kappa)$ in a Taylor series

$$\omega(\kappa) = \omega(k) + \omega'(k)\,(\kappa - k) + \tfrac{1}{2}\omega''(k)\,(\kappa - k)^2 + \ldots \quad , \tag{5.22}$$

we obtain

$$\psi(x, t) = e^{i[kx - \omega(k)t]}\int_{-\infty}^{\infty} A(\kappa)e^{i(\kappa-k)[x-\omega'(k)t] - \frac{i}{2}(\kappa-k)^2\omega''(k)t + \cdots}d\kappa\,. \tag{5.23}$$

As long as $$t \ll \frac{1}{(\Delta k)^2\,\omega''(k)} \tag{5.24}$$

the term $-\frac{1}{2}i(\kappa-k)^2\omega''(k)t$ can be neglected. Normally the higher terms will be even smaller, so that $u(x, t)$ will maintain the general form of Fig. 5.8b. The waves travel through the pattern with the phase velocity $k^{-1}\omega(k)$, while the envelope advances with the group velocity $\omega'(k)$. At later times, when condition (5.24) is breached, the packet will disperse.

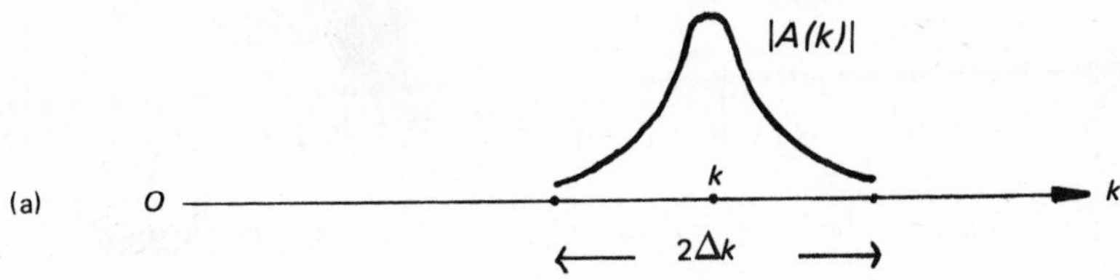

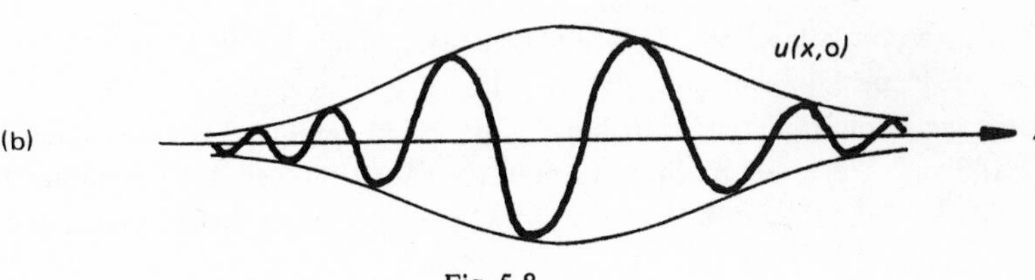

Fig. 5.8

5.7.3 Asymptotic behaviour

We now return to the superposition (5.21), with $A(\kappa)$ no longer restricted to a narrow band. Each component of this bundle of harmonic waves travels with a different velocity according to the dispersion law. We may ask how long it will take for the dispersion to mature, so that the component with a given wave number k is teased out of the bundle, and we may also seek the form of the resulting wave train in the neighbourhood of k. Therefore we set out to find an **asymptotic approximation** to $\psi(x, t)$ valid for large values of t.

Let $|A(\kappa)| = N(\kappa)$ and $\arg A(\kappa) = \epsilon(\kappa)$. Then the phase of the integrand in (5.21) is

$$\theta(\kappa) = \kappa x - \omega(\kappa)t + \epsilon(\kappa) \quad .$$

For t sufficiently large $e^{i\theta(\kappa)}$ oscillates rapidly over most of the domain of integration while $N(\kappa)$ changes slowly, and so the net contribution to the integral is very small. This holds provided that

$$t|\omega'(\kappa)| \gg \frac{|N'(\kappa)|}{N(\kappa)} \tag{5.25}$$

for all values of κ for which $N(\kappa)$ is not negligible.

The major contributions to the integral then arise from narrow bands of values of κ centred on the zeros of $\theta'(\kappa)$, where the phase varies relatively slowly. In the case where there is only one solution k of the equation $\theta'(\kappa) = 0$, let E be the corresponding narrow band around $\kappa = k$. Because E is a small domain and $N(\kappa)$ is slowly varying we may write $N(\kappa) \cong N(k)$ in E. This can be shown to be justifiable if

$$t|\omega''(k)| \gg \left\{\frac{N'(k)}{N(k)}\right\}^2 \quad \text{and } t|\omega''(k)| \gg |\epsilon''(k)| \quad . \tag{5.26}$$

If t satisfies the conditions (5.25) and (5.26) the expansion (5.23) leads to the asymptotic approximation

$$\psi(x, t) \sim A(k)\, e^{i[kx-\omega(k)t]} \int_E e^{-\frac{i}{2}(\kappa-k)^2\omega''(k)t}\, d\kappa, \; (t\to\infty) \quad . \tag{5.27}$$

This is Kelvin's **method of stationary phase**, which is closely related to the method of steepest descent (Jeffreys 1956).

Now we know that the contributions to the integral (5.23) from outside E are small, and therefore no important error is introduced by extending the integration (5.27) to the infinite domain. Let $\kappa = k + s$. Then the integral becomes

$$2\int_0^\infty e^{-i\alpha s^2}\, ds, \quad \text{where } \alpha = \frac{\omega''(k)t}{2} \quad ,$$

which can be evaluated by integrating $e^{-i\alpha s^2}$ round the contour bounded by

$$\arg s = 0, \; |s| = R(R\to\infty), \; \arg s = \mp\frac{\pi}{4} \text{ for } \alpha \gtrless 0 \quad .$$

The final result is

$$\psi(x, t) \sim A(k)\sqrt{\left(\frac{2\pi}{t|\omega''(k)|}\right)}\; e^{i[kx-\omega(k)t-\frac{\pi}{4}\operatorname{sgn}\omega''(k)]}, \; (t\to\infty) \tag{5.28}$$

where $\quad \operatorname{sgn}\xi = \pm 1 \text{ for } \xi \gtrless 0$

and $k(x, t)$ is the solution of $t\omega'(k) - \epsilon'(k) = x$. (5.29)

The asymptotic solution (5.28) has the form of (5.8). We conclude that, after a lapse of time satisfying (5.25) and (5.26), the dispersion will be sufficiently mature to merit description as in section 5.4.

If there are multiple solutions of (5.29), then (5.28) will be a sum of several terms, one for each of the solutions. This was discussed qualitatively in section 5.5. Note that the wave number k_D giving minimum group velocity satisfies $\omega''(k) = 0$. To obtain an asymptotic approximation valid at k_D it would be necessary to proceed to the next term of the Taylor series (5.22).

5.7.4 Energy transport

In most physical systems the mean energy density over a period of oscillation in a harmonic wave $ae^{i(kx-\omega t)}$ is of the form

$$\bar{\mathcal{E}} = |a|^2 f(k) \quad ,$$

where $f(k)$ depends on the structure of the system. For example, from (5.6), in the elastically anchored string

$$f(k) = \tfrac{1}{2}\rho(\beta^2 + c^2k^2) \quad .$$

In the slowly-varying wave train (5.28) the mean energy density

$$\bar{\mathcal{E}}(x, t) = \frac{f(k)\ |A(k)|^2 2\pi}{t|\omega''(k)|} \quad .$$

Consider the region $D(t)$ which is occupied at time t by waves with wave numbers between k and $k + \mathrm{d}k$. The length of this region is, from (5.29), and neglecting $\epsilon''(k)$ in accordance with (5.26),

$$\mathrm{d}x = t|\omega''(k)|\ \mathrm{d}k \quad . \tag{5.30}$$

Hence the energy in the region $D(t)$ is

$$\bar{\mathcal{E}}\mathrm{d}x = 2\pi\, f(k)|A(k)|^2\, \mathrm{d}k \quad .$$

This quantity is independent of t. Therefore, as the region $D(t)$ advances with velocity $U(k)$ the total energy within it remains constant, that is, the energy is transported at the group velocity. The length of $D(t)$ increases linearly with t, from (5.30), so that the energy is spread out and the amplitude diminishes as $t^{-1/2}$.

PROBLEMS

5.1 A uniform telegraph cable possesses inductance L, resistance R, capacitance C, and leakage conductance G, all per unit length. Show that the current $u(x, t)$ and the voltage $v(x, t)$ satisfy the equations

$$Lu_t + Ru + v_x = 0, \; Cv_t + Gv + u_x = 0, \qquad \text{(cf. equation 1.38)}$$

and deduce that $c^2u_{xx} = u_{tt} + 2\lambda u_t + \beta^2 u$ (the **equation of telegraphy**),

where

$$c = \frac{1}{\sqrt{(LC)}}, \; 2\lambda = \frac{R}{L} + \frac{G}{C}, \; \beta = \sqrt{\left(\frac{RG}{LC}\right)} .$$

Show that an undisturbed attenuated current signal

$$u = e^{-\alpha x} f(x - ct) \quad ,$$

where f is an arbitrary function, can be transmitted if $\lambda = \beta$, that is $RC = LG$ (called the **distortionless line** condition).

5.2 The dispersion relation for electromagnetic waves in a dielectric possessing one resonant frequency Ω is approximately

$$\frac{c^2k^2}{\omega^2} = 1 + \frac{\sigma^2}{\Omega^2 - \omega^2 - i\omega g} \quad ,$$

where c is the velocity of light in vacuo, σ is a constant and g is a constant depending on the dissipation due to molecular motion.
Assuming σ^2 to be small compared with the magnitude of the denominator, and putting $k = q + ip$, show that

the **refractive index** $\quad n = \dfrac{cq}{\omega} \cong 1 + \dfrac{\sigma^2}{2} \dfrac{\Omega^2 - \omega^2}{(\Omega^2 - \omega^2)^2 + \omega^2 g^2}$

and the **absorption coefficient** (attenuation/wavelength if $n \cong 1$)

$$\alpha = \frac{cp}{\omega} \cong \frac{\sigma^2}{2} \frac{\omega g}{(\Omega^2 - \omega^2)^2 + \omega^2 g^2} .$$

(The approximation is not valid when ω is close to Ω.)
Sketch the graphs of n and α against ω and note that near the resonance frequency Ω the absorption is high (opacity) and the dispersion is **anomalous** (increase in phase velocity with frequency). The reversal of the prismatic spectrum is only observable at the edge of the anomalous region because of the high absorption. It is observed in certain dyes.

Sketch also the $q - \omega$ graph.

5.3 A **ripple tank** uses water to simulate, in slow motion, the behaviour of non-dispersive sound waves. The depth of the water is chosen so as to make the dispersion curve (Fig. 5.6) as nearly straight as possible near the origin, that is, arranging for k_C to be zero. Show that the required depth is $\sqrt{(3\tau)}$. For water τ is 7.44 mm^2, giving about 5 mm for the optimum depth. Show that the ripple speed is then approximately 22 cm s^{-1}, and that the phase velocity V is constant to within 6% for wavelengths above 2 cm. [Expand V^2 in powers of kh.]

5.4 The elastically anchored string of section 5.3 is subjected to a transverse resisting force per unit length equal to $2\rho\lambda$ times the speed. Show that its equation of motion is the telegraphy equation $c^2 u_{xx} = u_{tt} + 2\lambda u_t + \beta^2 u$.

Show that the attenuated wave

$$u = \text{Re}\, e^{-px} e^{i(qx - \omega t)} (p,\ q,\ \omega \text{ positive})$$

is a solution provided that $c^2(p^2 - q^2) = \beta^2 - \omega^2$ and $c^2 pq = \lambda\omega$.

Show that the phase velocity (defined here as ωq^{-1}) lies between $\beta\lambda^{-1}c$ and c, and the attenuation index p between βc^{-1} and λc^{-1}.

Show that $\dfrac{\partial \mathcal{E}}{\partial t} + \dfrac{\partial \mathcal{F}}{\partial x} \neq 0$, and find the velocity W of mean energy transport in the given wave. Show that $W < c$ at all frequencies.

5.5 Show that the phase function in a homogeneous material must satisfy

$$\theta_{tt}\theta_{xx} = \theta_{xt}^2 \quad .$$

[The surface $z = \theta(x, t)$ is a **developable** surface in xtz-space.]

5.6 When the initial disturbance is confined to small values of x and t, the quantity $\epsilon'(k)$ in equation (5.29) is small, and the group lines for the dispersed waves are given by

$$\frac{d\omega}{dk} = \frac{x}{t} \quad .$$

Find for this case the functions $k(x, t)$, $\omega(x, t)$ and the phase $\theta(x, t)$, and sketch the group lines and phase lines, when

(i) $\omega(k) = \gamma k^2$, (ii) $\omega^2 = gk$, (iii) $\omega^2 = \beta^2 + c^2k^2$,

where γ, g, β and c are constants.

5.7 Show that the wavelength $\lambda(x, t)$ and the period of oscillation $T(x, t)$ in a state of mature dispersion satisfy the relation

$$U^2 \frac{\partial \lambda}{\partial x} + V^2 \frac{\partial T}{\partial t} = 0 \quad ,$$

where U is the group velocity and V is the phase velocity.

Consider the waves produced on deep water by an initial disturbance confined to small values of x and t. Assume $\omega^2 = gk$, that is, exclude long waves and ignore surface tension. Suppose that it is agreed that a given frequency is recognisably separated by the dispersion if the periods of two successive oscillations at a fixed point x differ by less than 5%. Show that the condition is satisfied if

$$gt^2 > 80\,\pi x \quad .$$

Find the least distance from the source at which waves of length λ satisfy the condition.

5.8 A piano string possesses some flexural rigidity, and the equation of motion of the stretched string is

$$c^2 u_{xx} - \beta^2 u_{xxxx} = u_{tt} \quad .$$

Find the dispersion law and show that the group velocity $U(k)$ and the phase velocity $V(k)$ satisfy $V < U < 2V$.

If the string is stretched over bridges at $x = 0$ and $x = l$, so that u and u_{xx} are zero at both ends, show that the normal frequencies are given by

$$\omega_n = \frac{n\pi}{l}\left(c^2 + \frac{\beta^2 n^2 \pi^2}{l^2}\right)^{\frac{1}{2}} \quad , \qquad n = 1, 2, \ldots \quad .$$

[This explains why a piano tuner 'stretches' the scale slightly.]

5.9 The transverse motion of a thin beam clamped at both ends is governed by the equation

$$u_{tt} + \gamma^2 u_{xxxx} = 0 \ (\gamma \text{ constant})$$

with the boundary conditions

$$u(0, t) = 0 = u(l, t), u_x(0, t) = 0 = u_x(l, t) \quad .$$

Show that the normal frequencies are given by $\omega = \gamma k^2$, where k is a solution of the equation $\cosh kl \cos kl = 1$, and that the normal modes are of the form

$$\frac{\cosh kx - \cos kx}{\cosh kl - \cos kl} - \frac{\sinh kx - \sin kx}{\sinh kl - \sin kl} .$$

5.10 Show that $u(x, t) = t^{-\frac{1}{2}} e^{ix^2/t}$ is a solution of the equation

$$u_{xxxx} + 16u_{tt} = 0 \quad ,$$

and find the functions $k(x, t)$ and $\omega(x, t)$ for dispersed waves, verifying that they satisfy the dispersion relation.

Express u as a Fourier integral $u(x, t) = \int_{-\infty}^{\infty} C(k)e^{i(kx-\omega t)}\mathrm{d}k$ and show that $C(k)$ is constant ('flat spectrum').

5.11 Given that
$$a(x) = \cos x, \qquad -\frac{\pi}{2} < x < \frac{\pi}{2},$$
$$= 0 \quad , \qquad |x| > \frac{\pi}{2},$$

express the solution of the problem

$$u_{tt} = c^2 u_{xx} - \beta^2 u, \; u(x, 0) = a(x), u_t(x, 0) = 0 \quad (c, \beta \text{ constant})$$

as an integral with respect to k.

5.12 It is sometimes convenient to regard $\omega(k)$ as an odd function. Show that the functions $A(k)$ and $B(k)$ defined in section 5.7 then satisfy the relation

$$A(-k) = \overline{A(k)} \;, \; B(-k) = \overline{B(k)} \quad ,$$

and that equation (5.19) can be written

$$u(x, t) = \mathrm{Re} \int_0^{\infty} [A(k)\, e^{-i\omega t} + B(k) e^{i\omega t}]\, e^{ikx} \mathrm{d}k \quad .$$

5.13 A **Gaussian** wave packet is one which has the form

$$\psi(x, t) = \int_{-\infty}^{\infty} e^{-\alpha(\kappa - k)^2} e^{i[\kappa x - \omega(\kappa)t]}\, \mathrm{d}\kappa \quad .$$

Show that $\psi(x, 0) = (\pi/\kappa)^{\frac{1}{2}} \; e^{ikx} e^{-x^2/4\alpha}$.

[The spectrum width Δk is usually taken as $2\alpha^{-1/2}$, because when $|\kappa - k| = \alpha^{-1/2}$ the spectrum amplitude is reduced to $1/e$ of its maximum. The packet width Δx is $2\alpha^{1/2}$. We note that $\Delta k \Delta x$ is constant for all such packets. When applied to de Broglie waves in quantum mechanics this is an expression of Heisenberg's uncertainty principle.]

Consider the case $\omega(k) = \gamma k^2$ (γ constant), which applies both to the beam flexure equation and to Schrödinger's equation for a free particle. Show that

$$|\psi(x, t)| = \left(\frac{\pi^2}{\alpha^2 + \gamma^2 t^2}\right)^{1/4} e^{-\alpha[x - U(k)t]^2/(4\alpha^2 + 4\gamma^2 t^2)} ,$$

where U is the group velocity.

The result shows that the wave packet spreads as t increases, while its amplitude diminishes, as we saw in section 5.7.4.

[In order to evaluate the integral, write

$$\kappa - k = \frac{i(x - Ut)}{2(\alpha + i\gamma t)} + z \quad .$$

By integrating round a suitable contour obtain an integral along the real axis, and then use a contour similar to that used to obtain (5.28).]

5.14 Show that the minimum phase velocity V_m on deep water (assume $\tanh kh = 1$) is found by taking $\tau k^2 = 1$, where τ is as given in Problem 5.3. Show that the wavelength is 1.7 cm and $V_m = 23$ cm s^{-1}. Show that for the minimum group velocity

$$\tau k^2 = \frac{2}{\sqrt{3}} - 1 \quad ,$$

giving a wavelength about 4 cm and $U_0 \cong 17$ cm s^{-1}. These will be the last waves seen leaving a disturbed region.

[It is easy to show that the 'deep-water' approximation is justified in this case if h exceeds 5 cm.]

CHAPTER 6

Discrete Systems

6.1 GENERAL INTRODUCTION

We now consider the propagation of waves in a discrete structure, that is, a structure consisting of a set of distinct units each separated by a finite distance and in which there is some form of interaction between each pair of units. Because we intend to investigate the passage of a harmonic wave through the structure, we naturally impose a further condition that the structure should exhibit some form of periodicity.

Possibly the simplest structure is the **one-dimensional monatomic lattice,** which consists of an infinite number of equal particles moving on a straight line under mutual interactions. The force of interaction may be provided in a variety of ways, but it is assumed to take the same form for any pair of particles, so that in the equilibrium position the particles are uniformly spaced, each at a distance a from its nearest neighbours. For example, the particles may be connected by light springs each of natural length a, the forces of interaction being the tensions in the springs. Another (equivalent) example is the transverse motion of particles attached to a light taut string; the effective interactions in this case are the transverse components of the tension.

Because the particles are collinear, a particular particle may have its equilibrium position chosen to represent the origin, and the equilibrium coordinate of the nth particle P_n may then be taken to be na. Denoting the displacement of P_n from its equilibrium position by $u_n(t)$, we can see that the lattice may execute motions of the form

$$u_n(t) = Ae^{i(kna-\omega t)}, \qquad n = 0, \pm 1, \pm 2 \ldots \tag{6.1}$$

where A, k and ω are constants. This can be termed a harmonic wave on the discrete lattice of frequency ω and wavelength $\lambda = 2\pi/|k|$. Substituting $k + 2r\pi/a$ for k, where r is an integer, does not alter $u_n(t)$, and so all such waves can be derived from values of k extending over an interval of length $2\pi/a$; for convenience we choose

$$-\pi < ka \leqslant \pi \quad . \tag{6.2}$$

Also, any two unequal values of k within the interval yield distinct sets of functions $u_n(t)$. Thus each possible wave in the lattice is uniquely characterised by a value of k in the interval. Using the convention (6.2), the wavelength λ assigned to any wave satisfies $\lambda \geqslant 2a$.

Each wavelength λ corresponds to an acceptable frequency ω and we shall find that the minimum wavelength $\lambda_0 = 2a$ corresponds to a critical, or cut-off, frequency ω_0 which is a characteristic of the lattice. In the monatomic lattice ω_0 is a maximum acceptable frequency and the system operates as a **low-pass filter** in that it will only transmit undistorted harmonic waves of frequency $\omega \leqslant \omega_0$. Waves of frequency $\omega > \omega_0$ are strongly attenuated.

Although all one-dimensional lattices exhibit some form of filtering they are not necessarily low-pass filters. In the diatomic lattice of section 6.3 the dispersion relation gives two acceptable frequencies corresponding to each wavelength and there are three critical frequencies ω_0, ω_1, ω_2 such that $0 < \omega_0 < \omega_1 < \omega_2$. In this case the system passes frequencies in the bands $0 < \omega \leqslant \omega_0$ and $\omega_1 \leqslant \omega \leqslant \omega_2$ and strongly attenuates frequencies outside these ranges. Such a lattice is called **a band-pass filter.**

6.2 A ONE-DIMENSIONAL MONATOMIC LATTICE

As an illustrative example we consider the particles, each of mass m, to be joined together by light elastic springs of natural length a and modulus ϵ. Let u_n denote the longitudinal displacement of the nth particle.

When a spring of natural length a and modulus ϵ is extended by a distance y the tension in the spring is equal to $\epsilon y/a$. So when the lattice is in a general displaced configuration the equation of motion of the nth particle is

$$m\ddot{u}_n = \frac{\epsilon}{a}(u_{n+1} - u_n) - \frac{\epsilon}{a}(u_n - u_{n-1})$$

$$= \frac{\epsilon}{a}(u_{n+1} - 2u_n + u_{n-1}) \quad . \tag{6.3}$$

For the lattice to transmit a harmonic wave we require u_n to have the form

$$u_n = e^{i(n\theta - \omega t)}, \text{ where } \theta = ka \quad .$$

Hence

$$-m\omega^2 = \frac{\epsilon}{a}(e^{i\theta} - 2 + e^{-i\theta}) = \frac{2\epsilon}{a}(\cos\theta - 1) \quad ,$$

and so

$$\omega = \omega_0 \sin(\theta/2) \quad , \tag{6.4}$$

where $\omega_0^2 = 4\epsilon/ma$.

This is the **dispersion relation** for the lattice as a wave medium. The lattice will therefore transmit harmonic waves with frequencies in the interval $0 \leqslant \omega \leqslant \omega_0$. Adopting the convention (6.2), we see that the cut-off frequency ω_0 corresponds to the minimum wavelength $2a$.

The phase velocity is

$$V = \frac{\omega_0}{k} \sin \frac{ka}{2} \tag{6.5}$$

and the group velocity is

$$U = \frac{\omega_0 a}{2} \cos \frac{ka}{2} \quad . \tag{6.6}$$

Fig. 6.1 shows the dispersion curve and the behaviour of $V(k)$ and $U(k)$ for positive values of k.

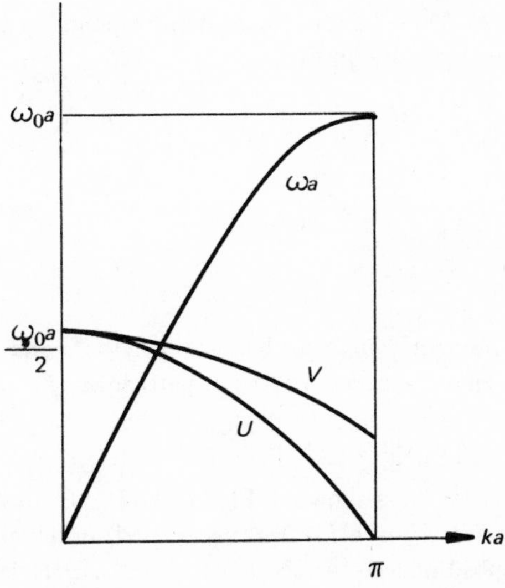

Fig. 6.1

When k is small the wavelength is large compared with the distance between adjacent particles and we see that V and U are both approximately constant and equal to $\frac{1}{2}\omega_0 a$. Hence the waves are propagated as if the lattice were a continuous string. As the wavelength decreases the phase velocity also decreases and approaches the value $\omega_0 a/\pi$ which corresponds to the minimum wavelength $2a$. It should be noted that at this wavelength V has no real meaning because $U = 0$, the particles are in unison, and there is effectively a standing wave.

Now consider what happens when we try to transmit a harmonic wave of frequency $\omega > \omega_0$. To do this put

$$\omega^2 = (1 + \tfrac{1}{2}\delta)\omega_0^2 \quad , \tag{6.7}$$

where $\delta > 0$. This satisfies the dispersion relation (6.4) if

$$\cos\theta = -(1+\delta) < -1 \quad . \tag{6.8}$$

Therefore θ must be complex. Let $\theta = \alpha + i\beta$. Then

$$\cos\alpha\cosh\beta - i\sin\alpha\sinh\beta = -(1+\delta) \quad . \tag{6.9}$$

Hence

$$\alpha = \pi \text{ and } \cosh\beta = 1 + \delta \quad . \tag{6.10}$$

The negative value of β may be rejected as it corresponds to a wave of increasing amplitude, which is physically unrealistic. So

$$\beta = \ln\{1 + \delta + \sqrt{(2\delta + \delta^2)}\} \tag{6.11}$$

Finally,

$$e^{ikx} = e^{-x\beta/a}\, e^{i\pi x/a} \quad , \tag{6.12}$$

so that a wave of frequency $\omega > \omega_0$ has wavelength $2a$ and is attenuated by a factor of $e^{-\beta}$ between each successive pair of particles.

6.3 A DIATOMIC LATTICE

The general diatomic lattice is shown in Fig. 6.2 and consists of a row of particles with alternating masses M, m $(M > m)$ free to perform longitudinal vibrations. In the equilibrium position a given mass M is connected to its right-hand neighbour by a light spring of natural length a and modulus ϵ and to its left-hand neighbour

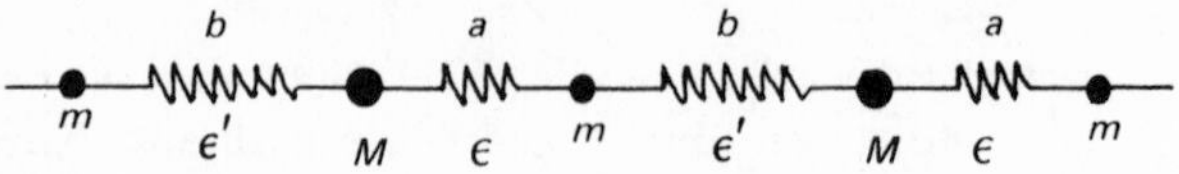

Fig. 6.2

by a spring of natural length b and modulus ϵ'. It is then possible to introduce a co-ordinate system such that in the equilibrium position there is a mass M at $x = (a+b)n$ and a mass m at $x = a + (a+b)n$ $(n = 0, \pm 1, \ldots)$. Finally in a general configuration let u_n, v_n denote the displacements from the equilibrium position of the masses M, m at $x = (a+b)n$ and $x = a+(a+b)n$ respectively. The equations of motion of the two types of particle are

$$M\ddot{u}_n = \alpha(v_n - u_n) + \beta(v_{n-1} - u_n) \tag{6.13}$$

and

$$m\ddot{v}_n = \alpha(u_n - v_n) + \beta(u_{n+1} - v_n) \quad , \tag{6.14}$$

where $\alpha = \epsilon/a$ and $\beta = \epsilon'/b$.

We now assume a wave solution to these equations of the form

$$u_n = Ae^{i[n(\theta+\phi)-\omega t]}, \; v_n = Be^{i[n(\theta+\phi)+\theta-\omega t]}, \tag{6.15}$$

where A and B are complex, $\theta = ka$ and $\phi = kb$, and k is restricted conventionally by the relation

$$-\pi < k(a+b) \leqslant \pi \quad . \tag{6.16}$$

This relation, like (6.2), ensures a unique correspondence between values of k and possible waves in the lattice. The minimum wavelength $2(a+b)$ is again twice the period of the lattice structure.

The first of the equations (6.15) represents a wave with frequency ω travelling through the particles of mass M, while the second represents a wave travelling through those of mass m with the same frequency but different amplitude and phase.

For solutions of this type we require that

$$\begin{aligned} -MA\omega^2 &= \alpha(Be^{i\theta} - A) + \beta(Be^{-i\phi} - A) \\ -mB\omega^2 &= \alpha(Ae^{-i\theta} - B) + \beta(Ae^{i\phi} - B) \quad . \end{aligned} \tag{6.17}$$

Hence

$$\begin{aligned} A(\alpha + \beta - M\omega^2) &= B(\alpha e^{i\theta} + \beta e^{-i\phi}) \\ A(\alpha e^{-i\theta} + \beta e^{i\phi}) &= B(\alpha + \beta - m\omega^2) \quad . \end{aligned} \tag{6.18}$$

For non-zero (A, B) the determinant of the coefficients must vanish. This gives

$$Mm\omega^4 - (M+m)(\alpha+\beta)\,\omega^2 + 4\alpha\beta\sin^2\tfrac{1}{2}(\theta+\phi) = 0 \quad , \tag{6.19}$$

which is a relation between ω and k in terms of the constants of the lattice.

This equation is satisfied by two positive values of ω^2, and hence for each value of k there will be two positive values of ω, and the dispersion curve will have two branches. So the system will, in general, have two frequency bands in which undistorted harmonic waves may be propagated. These are called the **passing bands** of the system.

The case $\alpha = \beta$ is of special interest because it arises in a variety of physical problems including the excitation of ions in simple crystals and the flow of current in electrical circuits. In this case (6.19) becomes

$$Mm\omega^4 - 2\alpha(M+m)\omega^2 + 4\alpha^2 \sin\theta = 0 \quad . \tag{6.20}$$

Hence

$$Mm\omega^2/\alpha = M+m \pm \sqrt{[(M+m)^2 - 4Mm\sin^2\theta]} \quad . \tag{6.21}$$

Fig. 6.3a shows the two branches of the dispersion curve. The passing bands are given by

$$0 \leqslant \omega \leqslant \omega_0 \text{ and } \omega_1 \leqslant \omega \leqslant \omega_2 \quad ,$$

where

$$\omega_0^2 = 2\alpha/M,\ \omega_1^2 = 2\alpha/m \text{ and } \omega_2^2 = 2\alpha(M+m)/Mm \quad .$$

The higher frequency band is often called the **optical branch** owing to the fact that when the theory is applied to the vibrations in a crystal lattice, the frequencies in this band are of the order of magnitude of infra-red frequencies. The lower frequency band is similarly called the **acoustical branch** because the frequencies in it roughly correspond to those of acoustic vibrations.

We now consider the motion of individual particles for frequencies within the two passing bands. The ratio of the amplitudes of the two waves passing through the masses M, m is

$$A/B = 2\alpha\cos\theta/(2\alpha - M\omega^2) \quad . \tag{6.22}$$

In the acoustical branch $A/B > 0$; we see from (6.15) that the masses M and m are in phase for long wavelengths (θ small), the phase difference θ increasing steadily from 0 and reaching $\pi/2$ at the cut-off frequency ω_0. In the optical branch $A/B < 0$ and the masses are in opposite phase at the highest frequency ω_2, with the phase difference decreasing from π to $\pi/2$ as the frequency falls from ω_2 to ω_1. An ionic crystal, in which ions of opposite polarity alternate, can thus be excited in the optical modes by an oscillating electric field.

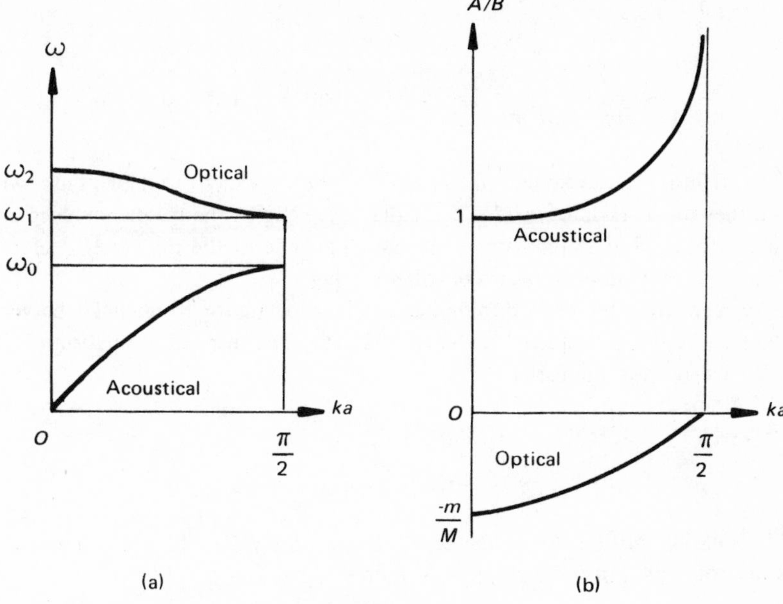

Fig. 6.3

The behaviour of A/B as k varies is shown in Fig. 6.3b. For small values of θ it is easily shown that in the acoustical branch

$$\omega^2 \approx \frac{2\alpha}{M+m}\,\theta^2 \text{ and } \frac{A}{B} \approx 1 + \frac{1}{2}\frac{M-m}{M+m}\,\theta^2 \quad , \tag{6.23}$$

whilst for values of θ near to $\pi/2$

$$\omega^2 \approx \frac{2\alpha}{M} - \frac{2\alpha}{M-m}\left(\frac{\pi}{2}-\theta\right)^2 \text{ and } \lim_{\theta\to\pi/2} A/B = \infty \quad . \tag{6.24}$$

Thus for an infinite wavelength the particles are in phase and have equal amplitudes. As the wavelength decreases to its minimum value the frequency increases to ω_0 and at the same time the amplitude of the masses m decreases until they are at rest with the masses M still vibrating.

In the optical branch, for small θ,

$$\omega^2 \approx 2\alpha\,\frac{M+m}{Mm} - \frac{2\alpha}{M+m}\,\theta^2 \text{ and } \frac{A}{B} \cong -\frac{m}{M}\left(1 - \frac{1}{2}\frac{M-m}{M+m}\,\theta^2\right) \quad , \tag{6.25}$$

and, when θ is near to $\pi/2$,

$$\omega^2 \approx \frac{2\alpha}{m} + \frac{2\alpha}{M-m}\left(\frac{\pi}{2} - \theta\right)^2 \text{, with } \lim_{\theta\to\pi/2} A/B = 0 \quad . \tag{6.26}$$

So for an infinite wavelength the masses M, m are exactly out of phase, with amplitudes in the ratio of m/M. For smaller wavelengths the frequency decreases from ω_2 to ω_1 and at the same time the amplitude of the masses M decreases until they are at rest with masses m still vibrating.

We now consider the stopping bands which occur for frequencies between ω_0 and ω_1 and for frequencies greater than ω_2. The dispersion relation (6.20) may be rewritten in the form

$$\omega^2\left(\frac{M+m}{2\alpha} - \omega^2\frac{Mm}{4\alpha^2}\right) = \sin^2\theta \quad , \tag{6.27}$$

and it is easily verified that when $\omega_0 < \omega < \omega_1$ the value of $\sin^2\theta > 1$ and so θ must be complex. Put $\theta = ka = \gamma + i\delta$, then

$$\sin\theta = \sin\gamma\cosh\delta + i\cos\gamma\sinh\delta \quad . \tag{6.28}$$

As this must be real we deduce that $\gamma = \pi/2$. Hence in this frequency band all waves must have the minimum wavelength, but as $\delta \neq 0$ the amplitudes are exponentially attenuated. For frequencies greater than ω_2 the value of $\sin^2\theta$ is negative and so $\sin\theta$ is pure imaginary. With the same notation as before this implies that $\gamma = 0$ and $\delta \neq 0$. So the waves are again attenuated along the lattice.

6.4 EQUIVALENT ELECTRIC CIRCUITS

It was shown in section 1.5 that analogies exist between mechanical and electrical systems whereby point masses correspond to inductances and interaction strengths correspond to capacitances. We now show that when an electric circuit consists of a set of given electrical units joined together in sequence then there may be an exact correspondence with the lattices previously considered.

6.4.1 A low-pass electric filter

Consider a set of identical units, each consisting of a capacitance C with one plate earthed and the other plate attached to an inductance L, joined together to form the circuit shown in Fig. 6.4.

The units in the circuit are numbered by selecting a given unit of capacitance and inductance as the zeroth, and then other units are naturally associated with the numbers $\pm 1, \pm 2, \ldots$. In unit number n let u_n be the current flowing through the inductance and let Q_n, V_n be the charge and potential on the capacitance. Applying Kirchhoff's laws to the circuit gives

$$V_{n-1} = L\dot{u}_n + V_n \tag{6.29}$$

and

$$u_n = u_{n+1} + \dot{Q}_n \quad . \tag{6.30}$$

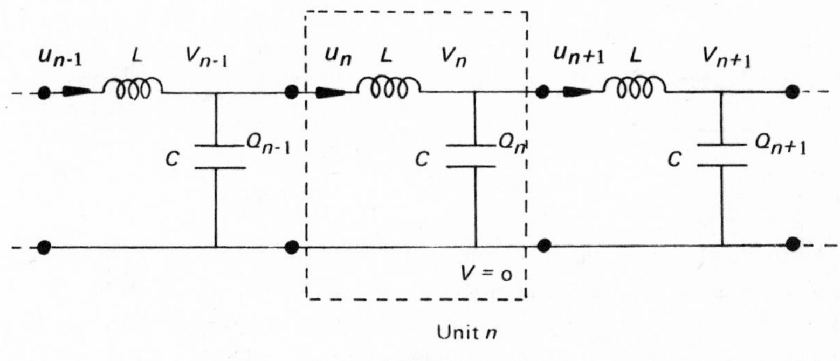

Fig. 6.4

After differentiating and using the relation

$$V_n = Q_n/C \tag{6.31}$$

we obtain

$$\ddot{u}_n = \frac{1}{LC}\,(u_{n+1} - 2u_n + u_{n-1}) \quad . \tag{6.32}$$

This is identical to the equation of motion of the lattice in section 6.2, equation (6.3), when ϵ/am is replaced by $1/LC$. Consequently we may deduce that the circuit is a low-pass filter in that it will transmit waves of frequency ω where $\omega \leqslant 2/\sqrt{(LC)}$, and waves above this critical frequency are strongly attenuated.

To obtain a high-pass electric filter we simply interchange the positions of the capacitance and inductance in each electrical unit (see Problem 6.6). The resulting circuit will then transmit waves of frequency $\omega \geqslant 1/\sqrt{(4LC)}$. It is also possible to construct a band-pass electric filter which will transmit waves in a frequency band

$$\omega_0 \leqslant \omega \leqslant \omega_1$$

where $\omega_0 \neq 0$ and ω_1 is finite. One way of doing this is to construct a circuit in which each electrical unit is identical to the one illustrated in Fig. 6.5. For such a circuit it is easily shown that

$$\omega_0 = 1/\sqrt{LC} \text{ and } \omega_1 = \sqrt{(5/LC)} \quad .$$

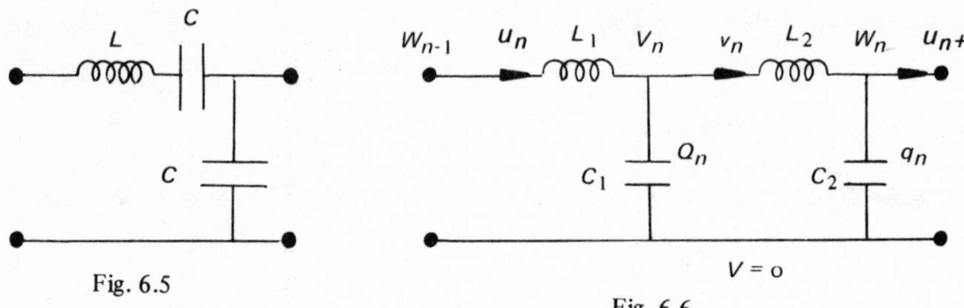

Fig. 6.5

Fig. 6.6

6.4.2 A double band-pass electric filter

A circuit which will transmit waves in two distinct frequency bands is obtained by a series of electrical units each identical to the one shown in Fig. 6.6. In unit number n let u_n, v_n be the currents flowing through inductances L_1, L_2 and let the potentials and charges on capacitances C_1, C_2 be V_n, W_n and Q_n, q_n respectively. Then the equations of the circuit are

$$\begin{aligned} u_n &= v_n + \dot{Q}_n,\ v_n = u_{n+1} + \dot{q}_n \quad , \\ V_n &= W_n + L_2\dot{v}_n,\ W_{n-1} = V_n + L_1\dot{u}_n \quad , \end{aligned} \tag{6.33}$$

together with $V_n = Q_n/C_1$, $W_n = q_n/C_2$.

Eliminating the potentials and charges gives

$$\begin{aligned} L_1\ddot{u}_n &= \frac{1}{C_1}(v_n - u_n) + \frac{1}{C_2}(v_{n-1} - u_n) \quad , \\ L_2\ddot{v}_n &= \frac{1}{C_1}(u_n - v_n) + \frac{1}{C_2}(u_{n+1} - v_n) \quad , \end{aligned} \tag{6.34}$$

and these equations may be identified with the equations of motion of the diatomic lattice (6.13) and (6.14) by replacing L_1, L_2 by M, m and $1/C_1$, $1/C_2$ by α, β respectively. Thus by direct analogy the circuit will transmit undistorted waves in two distinct frequency bands.

6.4.3 Resistance in a low-pass filter

One criticism of all the electric filters previously described is that they are unrealistic, owing to the fact that there is no resistance in the circuit. Whilst the introduction of resistance does produce slight changes in the properties of a filter, it does not change the basic characteristics. That is, a low (high or band)-pass filter is still a low (high or band)-pass filter after resistance is introduced. What does happen is that energy losses in the resistance produce attenuation, the amplitude of the wave diminishing as the wave progresses.

As an illustration, consider a circuit composed of electrical units each identical to the one shown in Fig. 6.7 where an inductance L and a resistance R are connected in series and shunted to earth by a capacitance C.

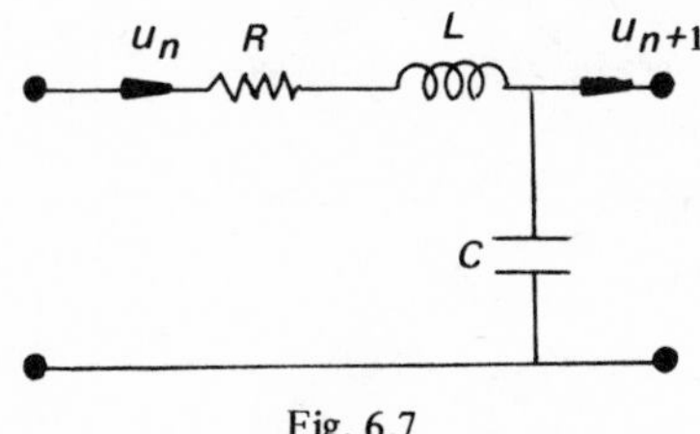

Fig. 6.7

In this case the equation for the current is found to be

$$LC\,\ddot{u}_n + RC\,\dot{u}_n = u_{n+1} - 2u_n + u_{n-1} \quad , \tag{6.35}$$

and for solutions of the form

$$u_n = e^{i(n\theta - \omega t)} \tag{6.36}$$

the dispersion relation is

$$LC\,\omega^2 + iRC\,\omega - 4\sin^2(\theta/2) = 0 \quad . \tag{6.37}$$

For real values of ω, θ will be complex, and waves of all frequencies will be attenuated. Fig. 6.8 shows the relations connecting ω with the wave number $\mathrm{Re}\theta$ and the attenuation index $\mathrm{Im}\,\theta$ in the case when the resistance is small, that is, $CR^2 \ll L$. In the absence of resistance $\omega = \omega_0|\sin(\theta/2)|$, where $\omega_0 = 2(LC)^{-1/2}$ is the cut-off frequency; when $R > 0$ the relation between ω and $\mathrm{Re}\,\theta$ follows the same curve but deviates from it when $\mathrm{Re}\,\theta$ is close to 0 or π. Frequencies above ω_0 can be transmitted, although they are heavily attenuated, as the $\mathrm{Im}\,\theta$ curve shows, but within the passing band there is only slight attenuation.

6.5 FINITE LATTICES

So far, we have only considered the propagation of progressive waves on infinite lattices. Whilst the theory may be applied to very long electrical systems it needs modifying slightly when applied to electric filters of finite length. Also in the study of the vibrations and absorption characteristics of an ionic crystal it is clear that the system is finite. Consequently we now investigate the way in which a finite lattice may support a wave form.

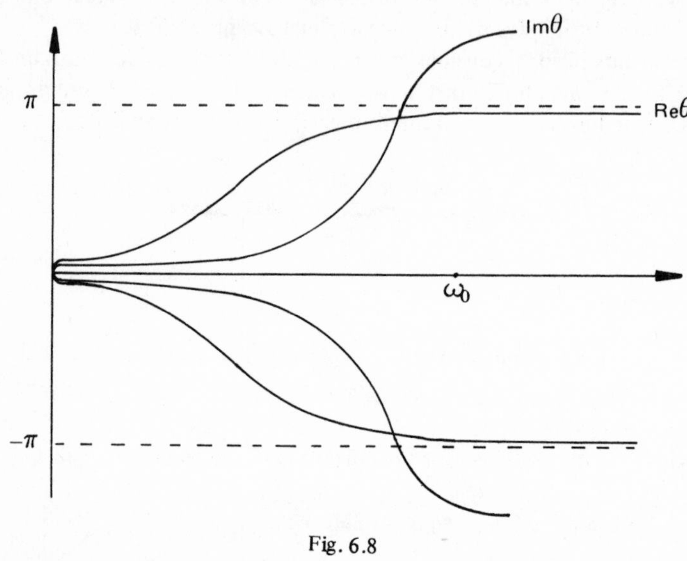

Fig. 6.8

Consider a lattice similar to the monatomic lattice of section 6.2, with particles at the points $x = na$ $(n = 1, 2, \ldots, N)$. The equation of motion of the nth particle, P_n, is (6.3) viz.

$$m\ddot{u}_n = \epsilon/a\,(u_{n+1} - 2u_n + u_{n-1}),\ (n = 2, 3, \ldots, N-1)\quad, \tag{6.38}$$

and the equations for u_1 and u_N will depend on the conditions imposed at $x = a$ and $x = Na$. It is a convenient device to assume that (6.38) also holds for $n = 1$ and $n = N$, and then to define the quantities u_0 and u_{N+1} in terms of $u_1, u_2, \ldots, u_N$ and time t so as to ensure that the end conditions are satisfied. For instance, when the lattice has **fixed ends**, with the points P_1 and P_N attached by springs to fixed points A at $x = 0$ and B at $x = (N+1)a$, the end conditions are $u_0 = u_{N+1} = 0$. Similarly, when the lattice has **free ends**, with P_1 and P_N not subject to any external forces, the end conditions are $u_0 = u_1$ and $u_N = u_{N+1}$.

As with continuous systems subject to boundary conditions, it is convenient to analyse the motion of a finite lattice in terms of the normal modes of oscillation; that is the standing wave solutions. We do this by substituting

$$u_n = U_n e^{-i\omega t} \tag{6.39}$$

in the equations (6.38), and we obtain the following recurrence relations for the U_n.

$$U_{n+1} + \left(\frac{m\omega^2 a}{\epsilon} - 2\right) U_n + U_{n-1} = 0, (n = 1, 2, \ldots, N) \quad ,(6.40)$$

with U_0 and U_{N+1} as linear functions of $U_1, \ldots, U_N$ given by the prescribed end conditions.

Substituting $U_n = e^{\pm in\theta}$ yields the dispersion relation

$$\omega^2 = \frac{4\epsilon}{ma} \sin^2(\theta/2) \quad , \tag{6.41}$$

which is the same as for the infinite lattice. So the normal solutions are of the form

$$U_n = Ae^{in\theta} + Be^{-in\theta}$$

where A, B are constants, and for a given frequency ω the corresponding value of θ satisfies (6.41). The end conditions involving U_0 and U_{N+1} must also be satisfied and these determine the ratio A/B together with a finite number of acceptable frequencies in the range $\omega^2 \leqslant 4\epsilon/ma$.

Consider, for example, the lattice with fixed ends. The end conditions are $U_0 = U_{N+1} = 0$. These give

$$A + B = 0 \text{ and } Ae^{i\theta} + Be^{-i\theta} = 0 \quad .$$

Hence $\sin(N+1)\theta = 0$ and $\theta = r\pi/(N+1)$, $(r = 0, 1, 2, \ldots)$. Using the periodicity of the sine and cosine functions we deduce that the lattice can execute longitudinal vibrations with frequency ω_r, where

$$\omega_r^2 = \left(\frac{4\epsilon}{ma}\right) \sin^2 \frac{r\pi}{2(N+1)} \quad , (r = 1, 2, \ldots, N) \quad ,$$

and in the rth mode the displacement u_n has the standing wave form

$$u_n = C \sin \frac{nr\pi}{N+1} \quad e^{-i\omega_r t} \quad ,$$

where C is a constant.

6.6 APPROACH TO A CONTINUOUS SYSTEM

When the distance of separation in the monatomic lattice becomes infinitely small the lattice clearly degenerates into a continuous system equivalent to that of a taut string of uniform line density. We now show that in this limiting case the equations of motion of the lattice reduce to the standard wave equation and that the behaviour of the lattice is identical to that of a uniform string, spring or rod, performing longitudinal vibrations.

The equation of motion of the particle of mass m at $x = na$ is

$$m\ddot{u}_n = (\epsilon/a)(u_{n+1} - 2u_n + u_{n-1}) \quad . \tag{6.42}$$

Now let $u(x, t)$ be a continuous function of two variables such that $u(na, t) = u_n$, for all integer n.
Then if, for the nth particle, $u(x, t) = u_n$ we have

$$u_{n-1} = u(x - a, t), u_{n+1} = u(x + a, t)$$

and

$$m\ddot{u}(x, t) = (\epsilon/a)\ [u(x + a, t) - 2u(x, t) + u(x - a, t)] \quad . \tag{6.43}$$

For a suitably continuous function

$$u(x + a, t) - 2u(x, t) + u(x - a, t) = a^2 u_{xx}(x, t) + O(a^3) \tag{6.44}$$

so that

$$m\ddot{u}(x, t) = \epsilon a \ u_{xx}(x, t) + O(a^3) \tag{6.45}$$

and in the limit as $a \to 0$,

$$\ddot{u}(x, t) = \lim_{a \to 0} \frac{\epsilon a}{m} u_{xx}(x, t) = \left(\frac{\epsilon}{\rho}\right) u_{xx}(x, t) \tag{6.46}$$

where ρ is the line density of the limiting continuous system. When a is small the dispersion equation of the lattice is

$$\omega^2 = \epsilon k^2 a/m \tag{6.47}$$

and in the limit as $a \to 0$

$$\omega^2 = (\epsilon/\rho)k^2 \quad . \tag{6.48}$$

Consequently as the monatomic lattice degenerates into a continuous system the equation of motion is

$$\ddot{u}(x, t) = c^2 u_{xx}(x, t) \tag{6.49}$$

where $c^2 = \epsilon/\rho$. All waves are propagated with the same speed c, and the lattice no longer acts as a filter because there is no cut-off frequency. Also all wavelengths $\lambda > 0$ correspond to distinct frequencies, and so there is no restriction on the wavelength of the motion. Hence the motion of the system is identical to that of a thin uniform rod executing longitudinal vibrations (cf. section 1.10.1).

PROBLEMS

6.1 A long uniform string is stretched to a tension T and carries a set of equal masses m, each separated from its immediate neighbours by a distance a. Show that the system is a low-pass filter in that it will transmit small *transverse* harmonic waves of the form $e^{i(kna-\omega t)}$ provided $\omega \leqslant \omega_0$, where $\omega_0^2 = 4T/am$, but that waves with $\omega > \omega_0$ are cut off.

6.2 A monatomic lattice consists of particles of mass m which in the equilibrium position are at $x = na$, $n = 0, \pm 1, \pm 2, \ldots$, and each particle is connected to its immediate neighbours by a spring of stiffness (force per unit extension) α. When each particle is subject to a further resistance equal to β times its speed, show that all longitudinal harmonic waves of the form $e^{i(kx-\omega t)}$ (k real) are damped with time, and that there are no oscillations if $\sin^2 \frac{1}{2}ka < \beta^2/16\alpha m$.

Show also that the damped waves are cut off when Re ω exceeds a certain critical frequency ω_1. Given that β is small, prove that

$$\omega_1 \approx (4\alpha/m)^{1/2}\,(1-\beta^2/32\alpha m) \quad .$$

6.3 In a monatomic lattice each mass m is connected to its immediate neighbours by springs of stiffness α and to its next-nearest neighbours by springs of stiffness β. Find the dispersion equation of the system given that in equilibrium adjacent masses are separated by a distance a. Deduce that the lattice is a low-pass filter and that for harmonic waves of frequency ω the passing band is $0 \leqslant \omega \leqslant \omega_0$, where

$$\omega_0^2 = 4\alpha/m \qquad \text{if } \beta < \alpha/4$$

and

$$\omega_0^2 = (\alpha+4\beta)^2/4m\beta \text{ if } \beta \geqslant \alpha/4 \quad .$$

6.4 A lattice consists of particles of mass m at $x = na$ ($n = -1, -2, \ldots$) and of mass M at $x = na$ ($n = 0, 1, 2, \ldots$). Each particle is connected to its nearest neighbours by a spring of stiffness α. Show that a harmonic wave $e^{i(kna-\omega t)}$

in the masses m may be partly reflected from and partly transmitted through $x = 0$. Show that the amplitude of the transmitted wave is $\sin ka \operatorname{cosec} \frac{1}{2}(K+k)a$ and its phase shift relative to the incident wave is $(K-k)a/2$, where K is the smallest positive solution of the equation $\sin(Ka/2) = \sqrt{(M/m)} \sin(ka/2)$.

6.5 A diatomic lattice consists of a row of alternating point masses $2m$, m free to move along the straight line Ox and connected together by springs. In the equilibrium position there is a mass $2m$ at $x = 2na$ and a mass m at $x = (2n+1)a$, ($n = 0, \pm 1, \pm 2, \ldots$). Each mass is connected to its immediate neighbours by a spring of stiffness α. Show that the lattice can carry a harmonic wave of frequency ω and propagation number k provided that $\cos\phi = z^2 - 3z + 1$, where $\phi = 2ka$ and $z = m\omega^2/\alpha$.

Show that in the stopping band ($1 < z < 2$) attenuated waves are propagated in which $\phi = \pi + i \cosh^{-1}(-1+3z-z^2)$.
In the case when $z = \frac{3}{2}$, show that the attenuation factor over the distance $2a$ is $\frac{1}{2}$.

6.6 Show that the electric circuit in Fig. 6.9 is a **high-pass filter** with a cut-off frequency of $\omega_0 = 1/\sqrt{(4LC)}$.

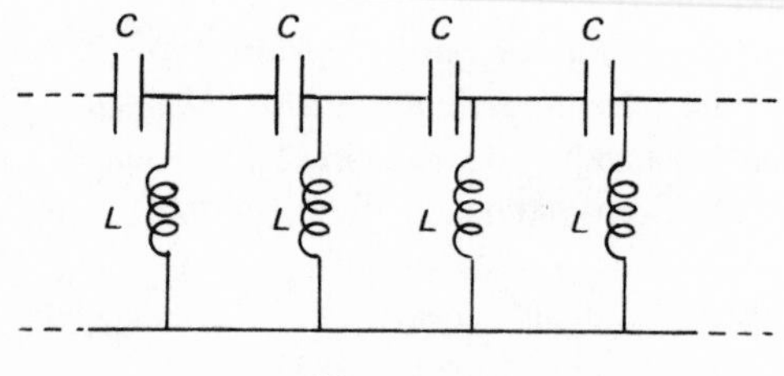

Fig. 6.9

6.7 A light taut string of length $4a$ has its ends fixed and carries particles of mass m at its points of quadrisection. Show that the frequencies of the normal modes of vibration are given by $\omega_1^2 = 2T/ma$, $\omega_2^2 = (2+\sqrt{2})T/ma$, $\omega_3^2 = (2-\sqrt{2})T/ma$ where T is the tension in the string.

Show also that the relative displacements of the masses in these modes is $1:0:-1$, $1:-\sqrt{2}:1$ and $1:\sqrt{2}:1$.

CHAPTER 7

The Fourier Method in Two Dimensions

7.1 THE HOMOGENEOUS WAVE EQUATION

In Chapters 3 and 4 when we solved the one-dimensional wave equation by the method of separation of variables we found that the normal modes of vibration were the eigenfunctions of a Sturm-Liouville system. When we apply the method to the two-dimensional equation we find that the normal modes are now the eigenfunctions of a *partial* differential equation with appropriate boundary conditions. Before solving any particular problems, we consider the general problem in order both to describe the procedure and to give some important results regarding the associated eigenvalue problem.

Consider the equation

$$c^2(u_{xx} + u_{yy}) = u_{tt}, (x, y) \epsilon\, G \quad , \tag{7.1}$$

where G is a region bounded by a simple closed contour Γ. We suppose the boundary condition to be

$$\alpha u + \beta \frac{\partial u}{\partial n} = 0, (x, y) \epsilon\, \Gamma \quad , \tag{7.2}$$

where α and β are non-negative functions defined on Γ and $\partial u/\partial n$ denotes the derivate in the direction of the outward-drawn normal to the curve Γ. Let the initial conditions be

$$u(x, y, 0) = a(x, y), u_t(x, y, 0) = b(x, y), (x, y) \epsilon\, G \quad . \tag{7.3}$$

As in Chapter 4 we separate the time variable by seeking normal solutions of the form

$$v(x, y)\,(C \cos\omega t + D \sin\omega t) \quad . \tag{7.4}$$

The normal mode $\nu(x, y)$ must satisfy

$$\nu_{xx} + \nu_{yy} + \lambda\nu = 0 \quad , \tag{7.5}$$

where $\lambda = \omega^2/c^2$, with

$$\alpha\,\nu + \beta\frac{\partial\nu}{\partial n} = 0 \quad \text{on } \Gamma \quad . \tag{7.6}$$

Just as in the one-dimensional case, it is found that non-zero solutions exist only for certain values of λ, the eigenvalues; the corresponding solutions $\nu(x, y)$ are the eigenfunctions, The proofs of the existence of eigenvalues and the associated theorems on completeness and series expansions are beyond the scope of this book (see, for example, Courant and Hilbert 1953). For our purposes it is sufficient to note the following results of the theory.

(a) There is an infinite set of linearly independent eigenfunctions $\nu_1(x, y)$, $\nu_2(x, y), \ldots$, with corresponding eigenvalues $\lambda_1, \lambda_2, \ldots$. Multiple eigenvalues may occur, in contrast to the one-dimensional case, and so there may be repetitions in the sequence $\{\lambda_n\}$.

(b) All the eigenvalues are non-negative.

(c) The eigenfunctions are mutually orthogonal in G, meaning that

$$\iint_G \nu_m(x, y)\,\nu_n(x, y)\,\mathrm{d}x\,\mathrm{d}y = 0 \text{ for } m \neq n \quad .$$

(see Problems 7.1 and 7.2)

(d) The eigenfunctions form a complete set. Any C^2 function $f(x, y)$ which satisfies the boundary condition (7.6) can be expanded in an absolutely and uniformly convergent series

$$f(x, y) = \sum_{n=1}^{\infty} F_n\,\nu_n(x, y) \quad . \tag{7.7}$$

(e) It follows from the orthogonality that the expansion coefficients F_n are given by

$$F_n \iint_G \nu_n^2(x, y)\,\mathrm{d}x\,\mathrm{d}y = \iint_G f(x, y)\,\nu_n(x, y)\,\mathrm{d}x\,\mathrm{d}y \quad . \tag{7.8}$$

(f) All the above properties apply to the wider class of problems pertaining to the equation $(p\nu_x)_x + (p\nu_y)_y + (\lambda r - q)\,\nu = 0$, where p_x, p_y, q, r are continuous and p, r are positive in G, the orthogonality integrals being weighted with the function $r(x, y)$, as in section 4.2.1.

We now form the superpositon

$$u(x, y, t) = \sum_{n=1}^{\infty} v_n(x, y)\,(C_n \cos\omega_n t + D_n \sin\omega_n t) \quad ,$$

where $\omega_n = c\sqrt{\lambda_n}$, and make u satisfy the initial conditions (7.3) by choosing C_n and D_n so that

$$a(x, y) = u(x, y, 0) = \sum_{n=1}^{\infty} C_n v_n(x, y)$$

and

$$b(x, y) = u_t(x, y, 0) = \sum_{n=1}^{\infty} \omega_n D_n v_n(x, y) \quad .$$

This can be achieved provided that a and b satisfy the conditions of the expansion theorem (d), and the required coefficients are given by

$$C_n \iint_G v_n^2(x, y)\, dx\, dy = \iint_G v_n(x, y)\, a(x, y)\, dx dy$$

and

$$\omega_n D_n \iint_G v_n^2(x, y)\, dx\, dy = \iint_G v_n(x, y)\, b(x, y)\, dx\, dy \quad .$$

7.2 THE INHOMOGENEOUS WAVE EQUATION

The solution of the inhomogeneous equation

$$u_{tt} = c^2(u_{xx} + u_{yy}) + p(x, y, t), (x, y) \in G, t > 0 \tag{7.9}$$

subject to the boundary condition (7.2) and the initial conditions (7.3), can be found by expansion in terms of the eigenfunctions $v_n(x, y)$ of the homogeneous equation following the method used in section 3.4.

Let $$u(x, y, t) = \sum_{n=1}^{\infty} v_n(x, y)\, S_n(t) \quad , \tag{7.10}$$

where the functions $S_n(t)$ are to be determined. This series satisfies the boundary condition. We expand the given functions p, a and b:

$$p(x, y, t) = \sum_{n=1}^{\infty} P_n(t)\, v_n(x, y) \quad , \tag{7.11}$$

$$a(x, y) = \sum_{n=1}^{\infty} A_n v_n(x, y) \text{ and } b(x, y) = \sum_{n=1}^{\infty} B_n v_n(x, y) \quad .$$

The coefficients $P_n(t)$, A_n and B_n are obtained by means of (7.8). Inserting (7.10) and (7.11) into (7.9) and using the uniqueness of the eigenfunction expansions we see that the functions $S_n(t)$ are determined by solving the ordinary differential equations

$$S_n''(t) + c^2\lambda_n S_n(t) = P_n(t)$$

with the initial conditions

$$S_n(0) = A_n,\ S_n'(0) = B_n,\ n = 1, 2, \ldots \quad .$$

7.3 THE VIBRATING MEMBRANE

A freely flexible stretched film is called a membrane. We consider a membrane which, in its equilibrium configuration, lies in the xy-plane and is subject to an **isotropic surface tension** $T(x, y)$ which is maintained by a static system of forces in the plane. This means that if we draw a line of length δs in the membrane then the force between the two parts of the membrane that are separated by the line has magnitude $T\delta s$ and is perpendicular to the line.

We now derive the equation of small transverse motion of the membrane. Assume that each point $P(x, y)$ of the membrane moves perpendicularly to the xy-plane, and let the displacement of P at time t be $z = u(x, y, t)$. By **small displacements** we mean that

$$u_x \ll 1 \text{ and } u_y \ll 1 \quad . \tag{7.12}$$

Fig. 7.1 shows the point P displaced to P_1, and the region S of the plane, bounded by the curve C, displaced to the surface S_1 bounded by the curve C_1.

The area of the surface S_1 is $\iint_S \sqrt{(1 + u_x{}^2 + u_y^2)}\, dx\, dy$. Neglecting u_x^2 and u_y^2 we deduce that the area of S_1 is equal to the area of S, and so the elastic deformation produces no additional tension. We may therefore, as in section 1.10.3, assume that the surface tension is always equal to its equilibrium value $T(x, y)$ and that the postulated small transverse motion is compatible with the force system.

At a point Q of C let $\mathbf{n}$ be the unit normal to C. Let $\mathbf{N}$ be the unit normal to the surface S_1 at the displaced point Q_1. Let Q' be a close neighbour of Q on C. The arc QQ' is displaced to Q_1Q_1' and, because of the small-displacement assumption (7.12), we may write $Q_1Q_1' \cong QQ' = \delta s$. The direction of Q_1Q_1' is $\mathbf{N} \times \mathbf{n}$. The force exerted on Q_1Q_1' by the part of the membrane outside S_1 is thus of magnitude $T\delta s$ and direction $\mathbf{n}_1 = (\mathbf{N} \times \mathbf{n}) \times \mathbf{N} = \mathbf{n} - (\mathbf{N} \,.\, \mathbf{n})\mathbf{N}$.

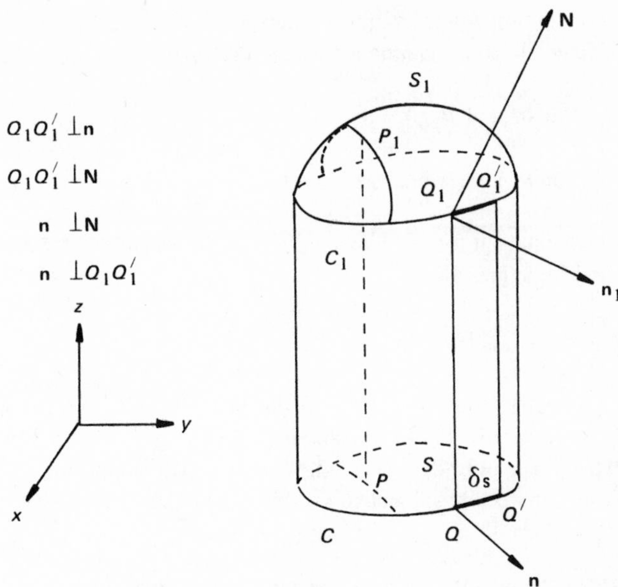

Fig. 7.1

Now $\mathbf{N}$ is the unit normal to the surface $z - u(x, y) = 0$, and so, to our order of approximation, its components are $(-u_x, -u_y, 1)$. Hence $\mathbf{N}.\mathbf{n} \cong -\mathbf{n}.\nabla u$, where ∇ is the two-dimensional gradient operator, giving $\mathbf{n}_1 = \mathbf{n} + (\mathbf{n}.\nabla u)\mathbf{N}$. Since ∇u is small, $\mathbf{n}_1$ is approximately a unit vector. The z-component of $\mathbf{n}_1$ is just $\mathbf{n}.\nabla u$, and so the transverse force exerted on $Q_1Q'_1$ is $(\mathbf{n}.\nabla u)\, T\delta s$. Hence the equation of small transverse motion of the region S of the membrane when it is subject to a transverse force density $g(x, y, t)$ is

$$\iint_S \rho u_{tt}\, \mathrm{d}x\mathrm{d}y = \oint_C (T\nabla u)\,.\,\mathbf{n}\, \mathrm{d}s + \iint_S g\, \mathrm{d}x\, \mathrm{d}y \quad ,$$

where $\rho(x, y)$ is the superficial density of the membrane. Applying the divergence theorem to the boundary integral we obtain

$$\iint_S \rho u_{tt}\, \mathrm{d}x\, \mathrm{d}y = \iint_S \left[\nabla.(T\nabla u) + g \right] \mathrm{d}x\, \mathrm{d}y \quad .$$

This relation holds for *every* region S. We may therefore equate the integrands, and obtain the differential equation governing the motion:

$$\rho u_{tt} = \nabla.(T\nabla u) + g \quad . \tag{7.13}$$

In the case of a uniform membrane, when ρ and T are constant, the equation of small transverse vibrations can be written in the form

$$u_{tt} = c^2(u_{xx} + u_{yy}) + p(x, y, t) \quad , \tag{7.14}$$

where $c^2 = T/\rho$ and $p(x, y, t) = \rho^{-1}g(x, y, t)$.

If there is no external force, (7.14) reduces to the equation for free vibrations of a uniform isotropic membrane,

$$u_{tt} = c^2(u_{xx} + u_{yy}) \quad . \tag{7.15}$$

The general boundary condition $\alpha u + \beta\ \partial u/\partial n = 0$ on Γ corresponds to an elastic attachment which exerts a transverse restoring force $\alpha\beta^{-1}Tu$ per unit length. When $\beta = 0$ the boundary is fixed. When $\alpha = 0$ the points on the boundary are free to move transversely subject to no transverse force.

7.4 FREE VIBRATIONS OF A RECTANGULAR MEMBRANE

7.4.1 The boundary and initial value problem

When we consider the small vibrations of a rectangular membrane which has its boundary fixed, and which in equilibrium occupies the region $G(0 \leqslant x \leqslant p, 0 \leqslant y \leqslant q)$, it is necessary to solve (7.15) subject to the boundary conditions

$$u(0, y, t) = u(p, y, t) = u(x, 0, t) = u(x, q, t) = 0 \quad . \tag{7.16}$$

We first seek normal solutions of the form

$$u(x, y, t) = v(x, y)\,(C \cos\omega t + D \sin\omega t) \quad .$$

Then

$$\begin{gathered} v_{xx} + v_{yy} + k^2 v = 0 \quad , \\ v(0, y) = v(p, y) = v(x, 0) = v(x, q) = 0 \quad , \end{gathered} \tag{7.17}$$

where $k^2 = \omega^2/c^2$.

To solve the eigenvalue problem given by (7.17) we again separate the variables. Substituting $X(x)\,Y(y)$ for $v(x, y)$ we obtain

$$\frac{X''(x)}{X(x)} + \frac{Y''(y)}{Y(y)} + k^2 = 0 \quad ,$$

so that

$$X''(x) + k_1^2 X(x) = 0; \; Y''(y) + k_2^2 Y(y) = 0 \quad , \tag{7.18}$$

where k_1^2, k_2^2 are constants such that

$$k^2 = k_1^2 + k_2^2 \quad . \tag{7.19}$$

The boundary conditions in (7.17) reduce to

$$X(0) = X(p) = 0, \quad Y(0) = Y(q) = 0 \quad . \tag{7.20}$$

The solutions of the two one-dimensional eigenvalue problems constituted by (7.18) and (7.20) are

$$X(x) = \sin k_1 x,\ Y(y) = \sin k_2 y, \text{ where}$$

$$k_1 = \frac{m\pi}{p},\ k_2 = \frac{n\pi}{q},\ (m, n = 1, 2, \ldots) \quad .$$

Consequently the eigenvalues of (7.17) are the quantities

$$k_{mn}^2 = \pi^2 \left(\frac{m^2}{p^2} + \frac{n^2}{q^2} \right) \quad ,$$

and the corresponding eigenfunctions are

$$v_{mn}(x, y) = \sin \frac{m\pi x}{p} \sin \frac{n\pi y}{q} \quad . \tag{7.21}$$

Hence, by superposition of normal solutions, we may write the general solution of (7.15) subject to (7.16) in the form

$$u(x, y, t) = \sum_{m=1}^{\infty} \sum_{n=1}^{\infty} (C_{mn} \cos ck_{mn}t + D_{mn} \sin ck_{mn}t) \sin \frac{m\pi x}{p} \sin \frac{n\pi y}{q} \quad .$$

The coefficients C_{mn}, D_{mn} are determined by the initial conditions

$$u(x, y, 0) = a(x, y),\ u_t(x, y, 0) = b(x, y) \quad .$$

We have $\quad a(x, y) = \sum_{m=1}^{\infty} \sum_{n=1}^{\infty} C_{mn} \sin \frac{m\pi x}{p} \sin \frac{n\pi y}{q}$

and $\quad b(x, y) = \sum_{m=1}^{\infty} \sum_{n=1}^{\infty} ck_{mn} D_{mn} \sin \frac{m\pi x}{p} \sin \frac{n\pi y}{q} \quad .$

The orthogonality property of the eigenfunctions is

$$\int_0^p \int_0^q v_{mn}(x, y)\, v_{\mu\nu}(x, y)\, dx\, dy = \begin{cases} \frac{1}{4}pq \text{ if } m=\mu \text{ and } n=\nu \\ 0 \text{ otherwise} \end{cases}$$

and so we deduce that

$$C_{mn} = \frac{4}{pq} \int_0^p \int_0^q a(x, y)\, v_{mn}(x, y)\, dx\, dy$$

and $$D_{mn} = \frac{4}{cpqk_{mn}} \int_0^p \int_0^q b(x, y)\, v_{mn}(x, y)\, dx\, dy \quad .$$

7.4.2 Normal mode patterns

The normal solutions of our problem tell us that, given suitable initial conditions, the membrane will vibrate in the normal mode $v_{mn}(x, y)$ with a natural harmonic frequency

$$\omega_{mn} = c\pi \sqrt{(m^2/p^2 + n^2/q^2)} \quad ,$$

and this illustrates an essential difference between a membrane and a string. For a string, each frequency of natural vibration corresponds to exactly one normal mode which has nodes at fixed points equally spaced along the string. However, for a membrane it is possible for one frequency of natural vibration to correspond to several independent normal modes and hence, as the system is linear, to any linear combination of these normal modes. The 'shape' of the membrane is thus not uniquely determined by the frequency. We illustrate this in Fig. 7.2 for the case of a square membrane of side p, in which $\omega_{mn} = (c\pi/p)\sqrt{(m^2 + n^2)}$ and $v_{mn}(x, y) = \sin(m\pi x/p) \sin(n\pi y/p)$. We indicate each mode shape by drawing the curves on which the displacement in the mode is always zero, called the **nodal lines**, and showing the relative signs of the displacement in the sub-regions bounded by the nodal lines. The nodal lines for the normal mode $v_{mn}(x, y)$ are simply lines parallel to the coordinate axes, four of which coincide with the boundary of the membrane. However, since $\omega_{mn} = \omega_{nm}$, both $v_{mn}(x, y)$ and $v_{nm}(x, y)$ have the same frequency of natural vibration and hence so does any linear combination of them. Thus for a frequency ω_{mn} the nodal lines are given by the equation

$$\alpha \sin \frac{m\pi x}{p} \sin \frac{n\pi y}{p} + \beta \sin \frac{n\pi x}{p} \sin \frac{m\pi y}{p} = 0 \quad ,$$

where α, β are arbitrary constants.

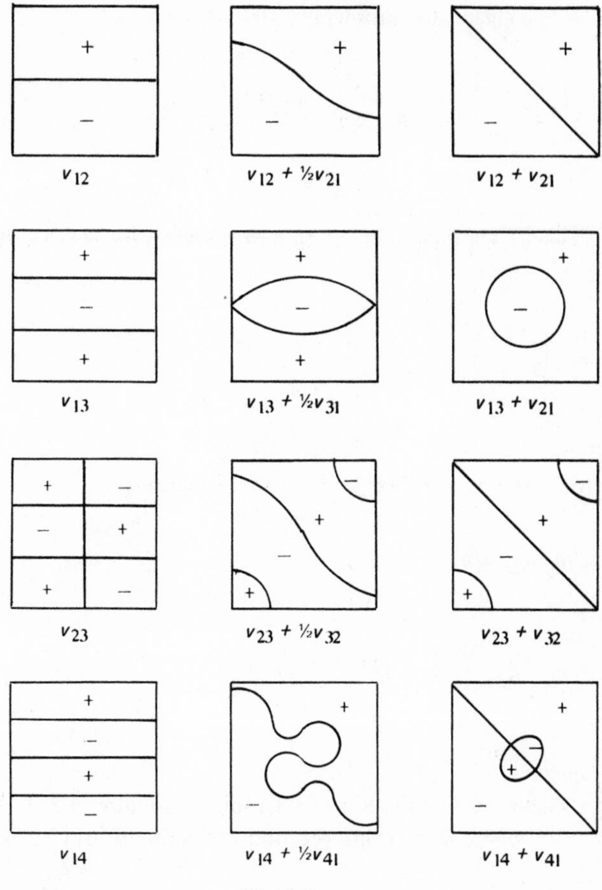

Fig. 7.2

7.5 FREE VIBRATIONS OF A CIRCULAR MEMBRANE

Let us now examine the vibrations of a circular membrane of radius p that is fixed at its edge. It is convenient to change to polar coordinates r, θ, when the wave equation assumes the form

$$\frac{1}{r}\frac{\partial}{\partial r}\left(r\frac{\partial u}{\partial r}\right)+\frac{1}{r^2}\frac{\partial^2 u}{\partial \theta^2}=\frac{1}{c^2}\frac{\partial^2 u}{\partial t^2}\;,0\leqslant r<p \tag{7.22}$$

and the boundary condition is

$$u(p,\theta,t)=0 \tag{7.23}$$

For normal solutions of the form

$$u(r,\theta,t)=U(r,\theta)\,(C\cos kct+D\sin kct) \tag{7.24}$$

we need to solve the eigenvalue problem for $U(r, \theta)$:

$$\frac{1}{r}\frac{\partial}{\partial r}\left(r\frac{\partial U}{\partial r}\right) + \frac{1}{r^2}\frac{\partial^2 U}{\partial \theta^2} + k^2 U = 0 \quad ,$$
$$U(p, \theta) = 0 \quad . \tag{7.25}$$

To find $U(r, \theta)$ we seek a separated solution of the form $R(r)\,\Theta(\theta)$. Substituting this into (7.25) we obtain

$$\frac{r\dfrac{d}{dr}\left(r\dfrac{dR}{dr}\right)}{R} + \frac{\Theta''}{\Theta} + k^2 r^2 = 0 \quad , \tag{7.26}$$

and so we arrive at the two ordinary differential equations

$$\Theta''(\theta) + \mu\Theta(\theta) = 0 \tag{7.27}$$

and
$$\frac{1}{r}\frac{d}{dr}\left(r\frac{dR}{dr}\right) + \left(k^2 - \frac{\mu}{r^2}\right)R = 0 \quad , \tag{7.28}$$

where μ is a constant.

For u to be a single-valued function of position we require $\Theta(\theta + 2\pi) = \Theta(\theta)$, and hence $\mu = n^2$, where n is an integer, and the solutions of (7.27) must be of the form

$$\Theta(\theta) = A \cos n\theta + B \sin n\theta \quad . \tag{7.29}$$

Then (7.28) is seen to be Bessel's equation of order n, of which the general solution is a linear combination of $J_n(kr)$ and $Y_n(kr)$. For u to be finite when $r = 0$ the coefficient of Y_n must be zero, and so we may take $R(r) = J_n(kr)$.

The normal solution satisfies the boundary condition (7.23) if $J_n(kp) = 0$. Consequently the eigenvalues of (7.25) are the quantities $k_{nm}^2 = (\zeta_{nm}/p)^2$, where ζ_{nm} $(m = 1, 2, \ldots)$ is a root of the equation $J_n(\zeta) = 0$ $(n = 0, 1, 2, \ldots)$. Each eigenvalue is double, since there is no restriction on A and B in (7.29), and the eigenfunctions U may be taken as

$$v_{nm}(r, \theta) = J_n(k_{nm}r) \cos n\theta \text{ and } w_{nm}(r,\theta) = J_n(k_{nm}r) \sin\theta \quad . \tag{7.30}$$

Hence the general solution $u(r, \theta, t)$ of (7.22) subject to (7.23) may be written in the form

$$u(r, \theta, t) = \sum_{n=0}^{\infty} \sum_{m=1}^{\infty} v_{nm}(r, \theta) \; [A_{nm} \cos ck_{nm}t + B_{nm} \sin ck_{nm}t]$$

$$+ \sum_{n=0}^{\infty} \sum_{m=1}^{\infty} w_{nm}(r,\theta) \; [E_{nm} \cos ck_{nm}t + F_{nm} \sin ck_{nm}t] \; . \quad (7.31)$$

The coefficients A_{nm}, B_{nm}, E_{nm} and F_{nm} may be determined from the initial conditions

$$u(r, \theta, 0) = a(r, \theta), \; u_t(r, \theta, 0) = b(r, \theta) \quad .$$

Setting $t = 0$ in (7.31) gives

$$a(r, \theta) = \sum_{n=0}^{\infty} \sum_{m=1}^{\infty} J_n(k_{nm}r) \; [A_{nm} \cos n\theta + E_{nm} \sin n\theta] \quad .$$

This is the Fourier series expansion of a periodic function in the interval $[0, 2\pi]$, and so the coefficients of $\cos n\theta$ and $\sin n\theta$ must be the Fourier coefficients. That is

$$\frac{1}{2\pi} \int_0^{2\pi} a(r, \theta) \mathrm{d}\, \theta = \sum_{m=1}^{\infty} A_{0m} \, J_0(k_{0m}r) \quad ,$$

$$\frac{1}{\pi} \int_0^{2\pi} a(r, \theta) \cos n\theta \; \mathrm{d}\theta = \sum_{m=1}^{\infty} A_{nm} J_n(k_{nm}r) \quad ,$$

$$\frac{1}{\pi} \int_0^{2\pi} a(r, \theta) \sin n\theta \; \mathrm{d}\theta = \sum_{m=1}^{\infty} E_{nm} J_n(k_{nm}r) \quad .$$

This reveals A_{nm} and E_{nm}, for any fixed n, as the coefficients in the expansion of certain functions of r in terms of the functions $J_n(k_{nm}r)$. From formula (13.18) of the *Appendix*, we see that if

$$f(r) = \sum_{m=1}^{\infty} f_m J_n(k_{nm}r) \; (0 \leqslant r \leqslant p)$$

then $$f_m p^2 J_{n+1}^2 (k_{nm}r) = 2 \int_0^p r f(r) \, J_n(k_{nm}r) \mathrm{d}r \quad .$$

Hence $$A_{nm} = \frac{2}{\pi p^2 J_{n+1}^2(k_{nm}p)} \int_0^p \int_0^{2\pi} ra(r, \theta) J_n(k_{nm}r) \cos n\theta \, dr \, d\theta \tag{7.32}$$

and $$E_{nm} = \frac{2}{\pi p^2 J_{n+1}^2(k_{nm}p)} \int_0^p \int_0^{2\pi} ra(r, \theta) J_n(k_{nm}r) \sin n\theta \, dr d\theta$$

A similar process determines the coefficients B_{nm}, F_{nm}; we simply replace $a(r, \theta)$ by $b(r, \theta)$ and A_{nm}, E_{nm} by $ck_{nm}\,B_{nm}$, $ck_{nm}\,F_{nm}$ in (7.32).

As each eigenvalue k_{nm}^2 corresponds to two independent eigenfunctions $v_{nm}(r, \theta)$ and $w_{nm}(r, \theta)$ we see that, as with the square membrane, the mode shape is not uniquely determined by the natural frequency of vibration. For a natural frequency $\omega_{nm} = ck_{nm}$ the nodal lines are determined by the equations

$$J_n(k_{nm}r) = 0 \ , \ \alpha\cos n\theta + \beta\sin n\theta = 0 \ ,$$

where α and β are arbitrary constants. These equations determine m concentric circles with radii $r_i = \dfrac{\zeta_{ni}}{\zeta_{nm}} p$, $(i = 1, 2, \ldots, m)$, and n diameters of the membrane given by $\theta_j = \dfrac{\psi + j\pi}{n}$, where $\psi = \tan^{-1}(-\beta/\alpha)$ $(j = 0, 1, \ldots, n-1)$. Fig. 7.3 illustrates a number of nodal lines.

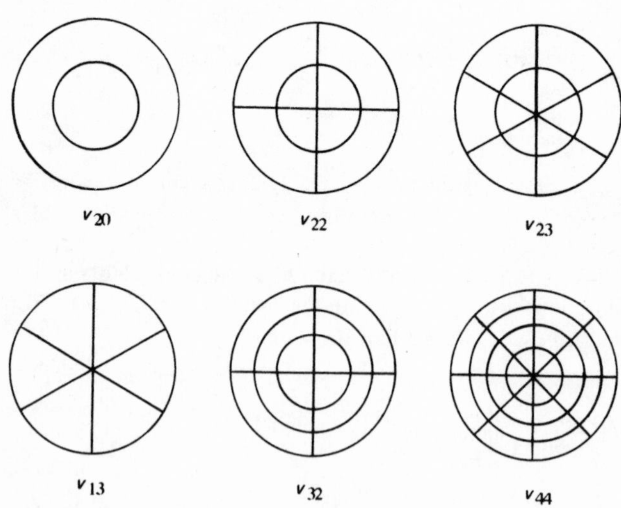

Fig. 7.3

PROBLEMS

7.1 Show that $\nabla.p\,(w\nabla v - v\nabla w) = w\nabla.\,p\nabla v - v\nabla.\,p\nabla w$, where ∇ is the two-dimensional gradient operator. (The formula is true generally in n dimensions.)

Hence show that if λ and μ are unequal eigenvalues with corresponding eigenfunctions v and w of the problem

$$\nabla.\,p\nabla v + (\lambda r - q)v = 0 \text{ in } G\,,$$

$\alpha v + \beta\,\partial v/\partial n = 0$ on Γ, the boundary of G,

then $$\iint_G rv\,w\,\mathrm{d}x\,\mathrm{d}y = 0 \quad .$$

7.2 If an eigenvalue possesses several linearly independent eigenfunctions it is necessary to form a mutually orthogonal set from them by suitable linear combinations in order to apply the expansion theorem. This can be achieved with any set of functions $\phi_j(x, y)$ $(j = 1, 2, \ldots, m)$ by the **Schmidt process.** Let $P_{ij} = \iint \phi_i\phi_j\mathrm{d}x\mathrm{d}y$. Then show that function $\psi_2 = \phi_2 + A\phi_1$ is orthogonal to ϕ_1 if $A = -P_{12}/P_{11}$. Now let $\psi_3 = \phi_3 + B\phi_1 + C\psi_2$, and find B and C in terms of the P_{ij} so that ψ_3 is orthogonal to both ϕ_1 and ψ_2. (The process may be continued until an orthogonal set of m functions is found.)

7.3 Given that α and β are small constants, find an expression for the transverse vibrations of the rectangular membrane of section 7.4 when

(a) $a(x, y) = \alpha \sin(\pi x/p)\sin(\pi y/q),\ b(x, y) = 0$.
(b) $a(x, y) = \alpha \sin(\pi x/p)\sin(\pi y/q),\ b(x, y) = \beta\sin(2\pi x/p)\sin(2\pi y/q)$.
(c) $a(x, y) = \alpha xy(p - x)(q - y),\ b(x, y) = 0$.

7.4 A uniform membrane is stretched over a rigid square frame of side p. The frame and membrane move with constant speed V in a direction normal to the membrane surface, which is plane. The frame is suddenly stopped at time $t = 0$, and remains at rest. Find an expression for the subsequent transverse vibrations of the membrane.

7.5 When a membrane vibrates in a medium which produces a resisting force proportional to the speed the equation of motion of free vibrations of the membrane is

$$u_{tt} + 2ku_t = c^2\nabla^2 u \quad .$$

Let the membrane be rectangular $(0 \leqslant x \leqslant p,\ 0 \leqslant y \leqslant q)$, with its edges fixed. Show that there are normal solutions

$$u_{mn} = e^{-kt}(A_{mn} \cos v_{mn}t + B_{mn} \sin v_{mn}t) \sin(m\pi x/p) \sin(n\pi y/q)$$

with characteristic frequencies

$$v_{mn} = \sqrt{(\omega_{mn}^2 - k^2)},\ m = 1, 2, \ldots; n = 1, 2, \ldots \quad ,$$

where the ω_{mn} are the characteristic frequencies of the undamped rectangular membrane and the A_{mn}, B_{mn} are arbitrary constants.

7.6 A uniform square membrane ($0 \leqslant x, y \leqslant p$) is fastened at its edges. At time $t = 0$ the membrane receives a blow which imparts an instantaneous transverse speed V to all points of a region $\mathfrak{D}$ of the membrane, whilst the remainder of the membrane remains instantaneously at rest. Given that $\mathfrak{D}$ is the region $|x - \frac{1}{2}p| < \epsilon$, $|y - \frac{1}{2}p| < \epsilon$, find the resulting transverse displacement of the membrane at a subsequent time t. Deduce an expression for the resulting displacement when the membrane is subject to an instantaneous impulse, of magnitude I, at its centre.

7.7 A uniform square membrane ($0 \leqslant x, y \leqslant p$) is fixed along its edges. Verify that there are normal modes with nodal line $x = y$, for the frequencies $(c\pi/p)\sqrt{5}$ and $(c\pi/p)\sqrt{10}$, where c is the wave velocity.

Find the next frequency, in increasing magnitude, possessing a mode with nodal line $x = y$. Show that these frequencies also have modes with nodal line $x + y = p$.

Sketch the nodal lines for the mode

$$\alpha \sin\theta \sin 2\phi + \beta \sin 2\theta \sin\phi,\ \theta = \pi x/p,\ \phi = \pi y/p \quad ,$$

in the cases $\alpha = 0$, $\alpha = \frac{1}{2}\beta$, $\alpha = \beta$, $\alpha = 2\beta$, $\beta = 0$.

7.8 Solve Problem 7.4 when the rigid frame is a circle of radius p.

7.9 Find an expression for the transverse vibrations of the circular membrane of section 7.5 when $a(r, \theta) = \alpha(p^2 - r^2)$ and $b(r, \theta) = 0$, where α is a small constant.

7.10 A uniform membrane has the shape of a circular annulus $q \leqslant r \leqslant p$ and is fixed at its edges $r = q$ and $r = p$. Its equation of small transverse motion is $c^2\nabla^2 u = u_{tt}$. Show that the frequencies of the circularly symmetric normal modes are given by $\omega = kc$, where k satisfies $J_0(kp)\,Y_0(kq) = J_0(kq)\,Y_0(kp)$.

7.11 Find an expression for the transverse vibrations of the circular membrane of section 7.5 when $a(r, \theta) = \alpha J_1(kr) \cos\theta$, $b(r, \theta) = \beta J_1(kr) \sin\theta$, where k is a root of the equation $J_1'(kp) = 0$ and α, β are small constants.

Show that any at time t there is a nodal line, and that the line rotates. Show further that the nodal line rotates with constant angular velocity when $kc|\alpha| = |\beta|$.

7.12 The circular membrane of section 7.5 is initially at rest in its equilibrium configuration. It is then subjected to a uniform pressure $f(t)$ applied over the region $r \leqslant q$, where $q < p$. Prove that at any subsequent time t the transverse displacement is given by

$$\frac{2q}{c\rho p^2} \sum_{n=1}^{\infty} \frac{J_1(qk_n) J_0(rk_n)}{k_n^2 J_1^2(pk_n)} \int_0^t f(s) \sin[ck_n(t-s)] \, ds,$$

where $k_n p$ is a root of the equation $J_0(\zeta) = 0$.

7.13 A homogeneous circular membrane of radius R is stretched to a tension T per unit length and its circumference is attached to a fixed support. The membrane lies on the surface of a liquid of density q, where it executes axi-symmetric vertical vibrations. Denote by $u(r, t)$ the vertical displacement of a point of the membrane. Noting that an element of area δA of the membrane is subjected to a hydrostatic force $-qgu\delta A$, show that the equation of the vibrations of the membrane is given by

$$u_{rr} + \frac{1}{r} u_r - b^2 u = \frac{1}{c^2} u_{tt},$$

where $b^2 = qg/T$, $c^2 = T/\sigma$ and σ is the surface density of the membrane.

Prove that the period of the fundamental vibration is

$$2\pi R \left(\frac{\sigma}{\zeta_1^2 T + qgR^2} \right)^{1/2},$$

where ζ_1 is the smallest positive root of the equation $J_0(\zeta) = 0$.

CHAPTER 8

Sound Waves in Fluids

8.1 EQUATIONS OF FLUID MOTION

The kinematical state of a fluid is defined by the **density field** $\rho(\mathbf{r}, t)$ and the **velocity field** $\mathbf{q}(\mathbf{r}, t)$.

The condition for **conservation of mass** is that the rate of increase of the mass of the fluid in any region Ω bounded by a closed surface Σ must equal the rate at which fluid is flowing inwards through Σ, that is

$$\frac{\mathrm{d}}{\mathrm{d}t} \int_{\Omega} \rho \,\mathrm{d}V = -\int_{\Sigma} \rho\, \mathbf{q}.\mathrm{d}\mathbf{S} \quad , \tag{8.1}$$

where $\mathrm{d}\mathbf{S}$ has the direction of the outward normal. Applying the divergence theorem, and demanding that the equation hold for every region, we arrive at the local form of the mass-conservation condition:

$$\frac{\partial \rho}{\partial t} + \mathbf{q} \,.\, \nabla \rho + \rho \operatorname{div} \mathbf{q} = 0 \quad , \tag{8.2}$$

which is known as the **equation of continuity**.

Let the position vector of a particle P moving with the fluid be $\mathbf{r}(t)$. As P moves from the position $\mathbf{r}$ at time t to the position $\mathbf{r} + \delta\mathbf{r}$ at time $t + \delta t$ it undergoes a velocity increment $\mathbf{q}(\mathbf{r} + \delta\mathbf{r}, t + \delta t) - \mathbf{q}(\mathbf{r}, t)$. The acceleration of P is therefore

$$\mathbf{f} = \frac{\partial \mathbf{q}}{\partial t} + \frac{\mathrm{d}\mathbf{r}}{\mathrm{d}t} .\nabla\, \mathbf{q} = \frac{\partial \mathbf{q}}{\partial t} + \mathbf{q}.\nabla \mathbf{q} \quad . \tag{8.3}$$

We make the following assumptions about the system of forces acting on the fluid.

Assumption 1 Energy dissipation, whether due to viscosity or to heat transfer, may be neglected. There are consequently no shear stresses, and so the internal stress can be derived from a single scalar field, the pressure $p(\mathbf{r}, t)$.

Assumption 2 The external forces are derived from a scalar field $W(\mathbf{r})$, the **potential energy per unit mass.**

The equation of motion of the fluid in the region Ω is

$$\int_{\Omega} \rho \mathbf{f} \, dV = -\int_{\Omega} \rho \nabla W \, dV - \int_{\Sigma} p d\mathbf{S} \quad .$$

By a corollary of the divergence theorem the last integral can be transformed to $\int_{\Omega} \nabla p \mathrm{d}V$. Now, the equation of motion holds for every region. Therefore, using (8.3), we arrive at the **Euler equation of motion**

$$\frac{\partial \mathbf{q}}{\partial t} + \mathbf{q} \cdot \nabla \mathbf{q} + \frac{1}{\rho} \nabla p + \nabla W = 0 \quad . \tag{8.4}$$

Assumption 3 The pressure is a known function $p(\rho)$ of the density (a relation called the equation of state). The rapidity of the density fluctuations in sound waves precludes significant heat flow, and so the **adiabatic equation of state** $p \propto \rho^{\gamma}$ is relevant. The constant γ depends on the gas; for air in normal conditions $\gamma = 1.4$.

The scalar equation (8.2) and the vector equation (8.4), together with appropriate initial and boundary conditions, determine the function $\rho(\mathbf{r}, t)$ and $\mathbf{q}(\mathbf{r}, t)$.

8.2 LINEARISATION OF THE EQUATIONS

We now consider **small** motions of the fluid, defined by the following approximations.

Approximation 1 The velocity is small compared with the instantaneous local speed of propagation of changes in any property of the fluid. For example, $|\mathbf{q}| \ll |\partial\rho/\partial t|/|\nabla\rho|$. Of course the condition may not be satisfied near points where $\partial\rho/\partial t = 0$, and $\nabla\rho \neq 0$ but we shall assume such regions to be too small to produce an appreciable effect on the solutions of the differential equations. The approximation enables us to replace (8.2) and (8.4) by

$$\frac{\partial \rho}{\partial t} + \rho \operatorname{div} \mathbf{q} = 0 \quad , \tag{8.5}$$

$$\frac{\partial \mathbf{q}}{\partial t} + \frac{\nabla p}{\rho} + \nabla W = 0 \quad . \tag{8.6}$$

Approximation 2 The **elastic deformation is small.** Let ρ_0 be the mean density over the whole region of interest. Define the **condensation** $s(\mathbf{r}, t)$ by the relation $\rho = \rho_0 (1 + s)$. We assume that $|s| \ll 1$ over the region of interest, and we may therefore write

$$\rho^{-1} \nabla p = \rho^{-1} p'(\rho) \nabla \rho = \rho^{-1} p'(\rho) \rho_0 \nabla s \cong c^2 \nabla s, \text{ where } c^2 = p'(\rho_0) \quad .$$

Equations (8.5) and (8.6) can now be replaced by their linearised approximate forms

$$\frac{\partial s}{\partial t} + \operatorname{div} \mathbf{q} = 0 \quad , \tag{8.7}$$

$$\frac{\partial \mathbf{q}}{\partial t} + \nabla(c^2 s + W) = 0 \quad . \tag{8.8}$$

It follows that $\mathbf{q} = \nabla f(\mathbf{r}, t) + \mathbf{A}(\mathbf{r})$, where f satisfies $\dfrac{\partial f}{\partial t} + c^2 s + W = 0$, and $\mathbf{A}(\mathbf{r})$ represents a steady flow. The time-varying part of the velocity field is therefore **irrotational.** Assuming that $\mathbf{A}(\mathbf{r})$ becomes constant when $|\mathbf{r}|$ is very large, we can apply Helmholtz's theorem, and write $\mathbf{A} = \nabla g + \operatorname{curl} \mathbf{B}$. Putting $f + g = \phi$, we obtain

$$\mathbf{q}(\mathbf{r}, t) = \mathbf{v}(\mathbf{r}, t) + \operatorname{curl} \mathbf{B}(\mathbf{r}) \quad , \tag{8.9}$$

where

$$\mathbf{v} = \nabla \phi \quad . \tag{8.10}$$

Substituting (8.9) into (8.7) and (8.8) gives

$$\frac{\partial s}{\partial t} + \operatorname{div} \mathbf{v} = 0 \quad , \tag{8.11}$$

$$\frac{\partial \mathbf{v}}{\partial t} + \nabla(c^2 s + W) = 0 \quad . \tag{8.12}$$

These equations govern the field $\mathbf{v}(\mathbf{r}, t)$. The steady-flow field curl $\mathbf{B}(\mathbf{r})$ makes no contribution to wave motion, and will henceforward be disregarded. We shall call $\mathbf{v}$ the velocity, and ϕ the **velocity potential.**

Integrating (8.12) we obtain $\dfrac{\partial \phi}{\partial t} + c^2 s + W = C(t)$, where $C(t)$ is an arbitrary function. We may take $C(t) = 0$, for any function of t can be added to ϕ without altering the velocity field $\nabla\phi$. Hence

$$\frac{\partial\phi}{\partial t} + c^2 s + W(\mathbf{r}) = 0 \quad . \tag{8.13}$$

From (8.10), (8.11) and (8.13) we deduce that

$$c^2 \nabla^2 \phi = \frac{\partial^2 \phi}{\partial t^2} \quad , \tag{8.14}$$

and so the velocity potential satisfies the classical wave equation. Usually the external force field $-\nabla W$ (e.g. gravity) is weak compared with the general level of the internal pressure forces, and so we may replace (8.13) by

$$\phi_t + c^2 s = 0 \quad . \tag{8.15}$$

These equations are valid only if the conditions of the linear approximation,

$$|s| \ll 1, \quad \mathbf{v} \,.\, \nabla s \ll s_t, \quad \mathbf{v} \,.\, \nabla \mathbf{v} \ll \mathbf{v}_t \tag{8.16}$$

are satisfied, and solutions of the equations must be tested for validity by substitution into these conditions. In what follows we shall assume that the conditions are satisfied.

8.3 WAVES IN TUBES

8.3.1 Cylindrical tubes with rigid walls

At a rigid wall the normal component of the velocity must vanish. The **boundary condition** is therefore $\partial\phi/\partial n = \mathbf{n}.\nabla\phi = 0$, where $\mathbf{n}$ is the unit normal. There is also a condition on the tangential component of velocity which arises from viscous forces and results in **absorption** of acoustic energy, but here we shall assume the absorption to be negligible.

Let the cross section of the tube be a region Σ bounded by a closed curve Γ in the xy-plane. Then the velocity potential satisfies

$$c^2 \nabla^2 \phi = \phi_{tt} \quad (x, y) \in \Sigma \quad , \tag{8.17}$$

$$\partial\phi/\partial n = 0 \quad (x, y) \in \Gamma \quad . \tag{8.18}$$

Let us seek solutions of the equation

$$\phi_{xx} + \phi_{yy} + \phi_{zz} = c^{-2}\phi_{tt}$$

in which the variables z and t are separated. The separation procedure leads us to solutions of the form $u(x, y)\, e^{i(kz - \omega t)}$, where $u(x, y)$ satisfies

$$u_{xx} + u_{yy} + (c^{-2}\omega^2 - k^2)\, u = 0 \text{ in } \Sigma, \text{ and } \partial u/\partial n = 0 \text{ on } \Gamma \quad . \tag{8.19}$$

This is an eigenvalue problem similar to the problem of a vibrating membrane with a 'free' boundary (equation 7.15). All the eigenvalues are non-negative (see section 7.1), and zero is certainly an eigenvalue, with $u =$ constant. Therefore we may write

$$c^{-2}\omega^2 - k^2 = \beta_j^2 \quad , \tag{8.20}$$

where $0 = \beta_0 < \beta_1 \leqslant \beta_2 \leqslant \ldots$.
Each eigenfunction $u_j(x, y)$ provides a **mode of propagation** which travels along the tube at a phase speed $V = \omega/k$ determined by the **dispersion relation**

$$\omega^2 = c^2k^2 + c^2\beta_j^2 \quad . \tag{8.21}$$

The group speed is $\mathrm{d}\omega/\mathrm{d}k = c^2k/\omega \leqslant c$, so c is the maximum signal speed. In the lowest mode $\omega = ck$ and $\phi \propto e^{ik(z-ct)}$. This is a plane wave, also longitudinal because $\nabla\phi$ is in the direction of propagation, and there is no dispersion, because all frequencies propagate at speed c.

If the tube is closed by rigid plane ends at $x = 0$ and $x = L$ it becomes a **cavity resonator.** The condition $\partial\phi/\partial n = 0$ on the end boundaries then restricts the separated solutions to the normal vibrations $\phi = u_j(x, y)\cos(n\pi z/L)e^{-i\omega_{jn}t}$, with normal frequencies given by

$$\omega_{jn}^2 = \left(\beta_j^2 + \frac{n^2\pi^2}{L^2}\right) c^2 \quad (j, n = 0, 1, 2, \ldots; \text{not both zero}). \tag{8.22}$$

8.3.2 Longitudinal waves in narrow tubes

Waves of frequency less than $c\beta_1$ can be propagated only in the longitudinal mode, as (8.20) shows. Now β_1^{-1} is of the order of magnitude of the tube 'width' (say, the largest diameter). For example if Γ is a rectangle with sides $a > b$, then $\beta_1^{-1} = \pi^{-1}a$ (see problem 8.2). So, if $\omega < c\beta_1$ the tube width is less than the wavelength, and for such waves the tube is **narrow**.

For this essentially one-dimensional motion it is convenient to take the x-axis parallel to the tube. The velocity potential is then a function $\phi(x, t)$ which must satisfy the equation

$$c^2\phi_{xx} = \phi_{tt} \tag{8.23}$$

Consider a wave of arbitrary shape, $\phi = f(x - ct)$, travelling along the tube.

Let $p = p_0 + \tilde{p}$, where $p_0 = p(\rho_0)$, the equilibrium pressure, and $\tilde{p}$ is the **excess pressure.** Then

$$\tilde{p} \cong \frac{\mathrm{d}p}{\mathrm{d}\rho} \rho_0 s = \rho_0 c^2 s = -\rho_0 \phi_t = \rho_0 c f' \quad . \tag{8.24}$$

We define the **characteristic impedance** Z of the tube as the ratio of the excess pressure to the rate of flow. If S is the area of the cross-section,

$$Z = \frac{\tilde{p}}{Sv} = \frac{-\rho_0 \phi_t}{S\phi_x} = \frac{\rho_0 c}{S} \quad . \tag{8.25}$$

The **intensity** of the sound in the plane wave is defined as

$$I = \tilde{p}v = -\rho_0 \phi_t \phi_x = \rho_0 c f'^2 \quad . \tag{8.26}$$

The **energy flux** in the tube is

$$\mathcal{F} = IS = \tilde{p}Sv = -\rho_0 \phi_t \phi_x S = \rho_0 c S f'^2 = \frac{\tilde{p}^2}{Z} \quad . \tag{8.27}$$

The **closed end** boundary condition is, as we have already seen,

$$\phi_x = 0 \quad . \tag{8.28}$$

At an **open end** there is no restriction on the velocity, but the intensity radiated in all directions rapidly falls off with distance, so that near the end the pressure has the equilibrium value p_0. Except for a small correction to allow for the radiation, which we shall ignore, the effective boundary condition is therefore $\tilde{p} = 0$, that is

$$\phi_t = 0, \text{ (all } t\text{) or, equivalently, } \phi = 0 \quad . \tag{8.29}$$

8.3.3 Junction of two tubes

Let two tubes T_1, T_2 with cross-section S_1, S_2 be joined at $x = 0$ (Fig. 8.1). We assume the change in area to be abrupt, meaning that the thickness d of the

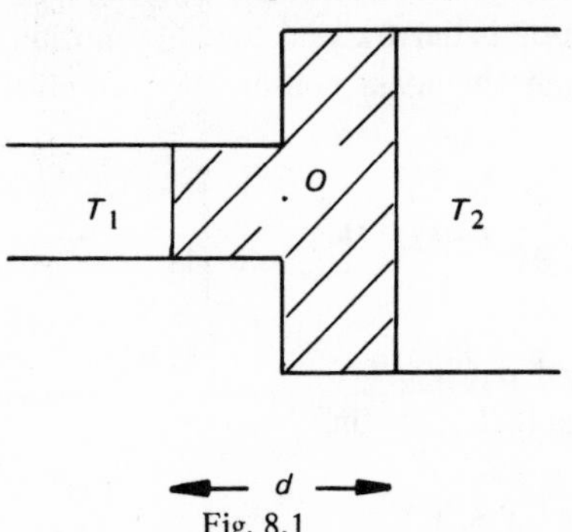

Fig. 8.1

shaded region in which the motion is not wholly longitudinal is small compared with the wavelength. The properties of the fluid in the wave therefore change very little over a distance d, and so we may take both boundaries of the region to be at $x = 0$. By considering the motion of the small volume of fluid in the shaded region we obtain the appropriate boundary conditions:

From the equation of motion $p(-0) = p(+0)$. (8.30)

From conservation of mass $S_1 v(-0) = S_2 v(+0)$. (8.31)

Hence, as in (8.29),

$$\phi(-0) = \phi(+0)$$

and

$$S_1\phi_x(-0) = S_2\phi_x(+0) \quad . \tag{8.32}$$

If an incident wave $\phi(x, t) = f(x - ct)$ approaches the junction along the tube T_1 it will be partly reflected and partly transmitted. Proceeding as in section 2.5.1 we write

$$\phi(x, t) = \begin{cases} f(x-ct) + g(x+ct) & x<0 \\ h(x-ct) & x>0 \end{cases} ,$$

and find the functions g and h in terms of f by using the boundary conditions (8.32). The results are

$$f(\zeta) : g(\zeta) : h(\zeta) = (S_1 + S_2) : (S_1 - S_2) : 2S_1 \quad .$$

As expected, there is no reflection if $S_1 = S_2$. From (8.27) we see that the ratios of the energy flux in the three waves is $(S_1 + S_2)^2$: $(S_1 - S_2)^2$: $4S_1S_2$. There is total reflection if $S_2 = 0$ (closed end) or $S_2 = \infty$ (open end).

8.3.4 Longitudinal waves in non-uniform tubes

We now consider a tube with continuously varying cross-section $S(x)$ (Fig. 8.2). We assume that the tube is narrow and that the motion is described by a plane wave $\phi(x, t)$. Applying the mass conservation condition (8.1) to the shaded region we obtain

$$\frac{\partial}{\partial t}(\rho S) + \frac{\partial}{\partial x}(\rho S v) = 0 \quad . \tag{8.33}$$

Linearising, $S\rho_0 s_t + \rho_0(Sv)_x = 0$.
Substituting for s using (8.15) we find

$$\phi_{tt} = c^2 S^{-1}(S\,\phi_x)_x \quad . \tag{8.34}$$

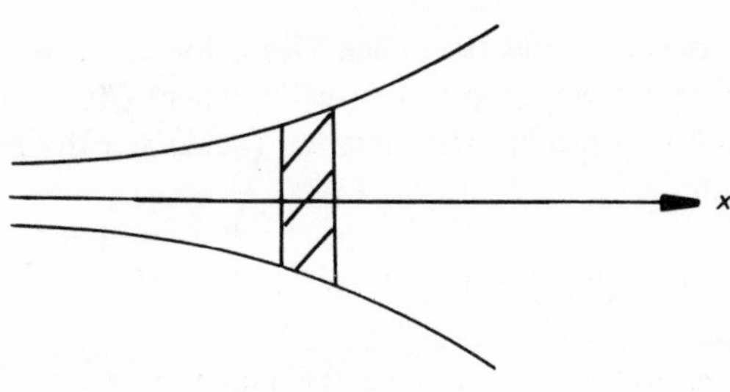

Fig. 8.2

If $S(x)$ is given, this equation can be solved by separation of variables. For example, the normal vibrations of the fluid in a tube of length L with the end $x = 0$ open and the end $x = L$ closed may be found by substituting

$$\phi = \Phi(x)e^{-i\omega t}$$

and solving the Sturm-Liouville problem:

$$c^2(S\Phi')' + \omega^2 S\Phi = 0 \tag{8.35}$$

subject to $\Phi(0) = 0, \Phi'(L) = 0.$

8.3.5 Horns

We have noted that the open end of a narrow tube reflects nearly all the wave energy back along the tube. The open end of a wide tube (of width exceeding the wavelength) exhibits the opposite behaviour: most of the energy escapes from the aperture and forms a forwardly directed beam. We shall not prove these results, which are fully discussed in books on acoustics (e.g. Morse 1948).

A tube of increasing cross-section which converts a narrow aperture into a wide one with minimum loss of wave energy acts as a **horn.** It permits the efficient emission of sound from a narrow tube. To design a horn we seek a solution of (8.34) of the form

$$\phi = A(x)\, e^{i(kx-\omega t)} \quad , \tag{8.36}$$

where $A(x)$ is a slowly varying function. The energy flux, from (8.27) is

$$\mathcal{F} = -\rho \text{ Re } \phi_t \text{ Re}\phi_x S \quad .$$

Averaging over one period of oscillation we obtain the mean energy flux

$$\overline{\mathcal{F}} = \tfrac{1}{2}\rho\, \omega k A^2 S \quad .$$

In order to prevent reflection back along the horn we require $\bar{\mathfrak{F}}$ to be constant, and therefore we choose $A = aQ^{-1}$, where $Q(x) = \sqrt{S}$ is the profile of the horn, and a is constant. The equation (8.34) and the proposed solution (8.36) then become

$$Q^2\phi_{tt} = c^2(Q^2\phi_x)_x \text{ and } \phi = aQ^{-1}\,e^{i(kx-\omega t)} \quad .$$

Substituting the expression for ϕ into the equation we find that there is a solution if

$$Q'' = \alpha^2 Q, \text{ where } \alpha = \sqrt{(c^{-2}\omega^2 - k^2)}$$

If $\alpha = 0$ the horn is conical, and

$$Q(x) = Q(0) + Q'(0)x \quad .$$

For $\alpha > 0$ the profile is

$$Q(x) = Q(0)\cosh\alpha x + \alpha^{-1}Q'(0)\sinh\alpha x \quad .$$

The constant $Q(0)$ is determined by the size of the aperture to which the horn is to be fitted. The values of $Q'(0)$ and α are available to be chosen, depending on the performance required of the horn. The conical horn transmits all frequencies at the same speed c, but is inefficient because of the reflection produced by the abrupt change in $Q'(x)$ at the narrow end. The disadvantage is less marked in the exponential horn $Q(x) = Q(0)\,e^{\alpha x}$, and is absent in the catenoidal horn $Q(x) = Q(0)\cosh\alpha x$. These horns exhibit dispersion through the relation $\omega^2 = c^2(\alpha^2 + k^2)$, and cannot transmit below the **cut-off** frequency $\omega_0 = c\alpha$. However, for frequencies well above ω_0 they are efficient, and the dispersion is not noticeable.

8.3.6 Elastic-walled tubes

Consider a tube whose cross-section in equilibrium is S_0 (constant), but is in general dependent on the pressure. Equation (8.33) may be written

$$\left(S\frac{d\rho}{dp} + \rho\frac{dS}{dp}\right)\frac{\partial p}{\partial t} + \frac{\partial}{\partial x}(\rho S v) = 0 \quad .$$

Linearising, assuming the fluctuations in S and ρ to be small, and remembering that $p_t = c^2\rho_0 s_t = -\rho_0\phi_{tt}$, we obtain

$$\left(\frac{d\rho}{dp} + \frac{\rho_0}{S_0}\frac{dS}{dp}\right)\phi_{tt} = \phi_{xx} \quad , \qquad (8.37)$$

the derivatives in the bracket being evaluated at the equilibrium pressure p_0. Hence longitudinal waves travel with speed V, where

$$\frac{1}{V^2} = \left(\frac{d\rho}{dp}\right)_0 + \frac{\rho_0}{S_0}\left(\frac{dS}{dp}\right)_0 = \frac{1}{c^2} + \frac{\rho_0}{S_0}\left(\frac{dS}{dp}\right)_0 . \tag{8.38}$$

If the walls are nearly rigid the last term is small and $V \cong c$. If the walls are very weak the last term dominates, and V becomes much lower than c. The fluid may be regarded as almost incompressible and the system becomes analogous to a channel containing liquid with a free surface (see Chapter 10).

8.4 CYLINDRICAL WAVES

The wave equation in cylindrical polar coordinates (r, θ, z) is

$$\frac{1}{r}(r\phi_r)_r + \frac{1}{r^2}\phi_{\theta\theta} + \phi_{zz} = \frac{1}{c^2}\phi_{tt} . \tag{8.39}$$

This equation is the same as the circular membrane equation (7.22), with the addition of the term ϕ_{zz}. Separated solutions are found by following the derivation in section 7.5. A typical solution is

$$[AJ_n(\beta r) + BY_n(\beta r)]\,(C\cos n\theta + D\sin n\theta)(E\cos\gamma z + F\sin\gamma z)(G\cos\omega t + H\sin\omega t),$$

where $c^2\beta^2 = \omega^2 - c^2\gamma^2$. (8.40)

There are three disposable constants: n, γ and ω.

8.4.1 Interior of a cylinder

To ensure continuity at $r = 0$, we must have $B = 0$.

If the cylinder has rigid walls $r = a$, then the radial velocity must be zero at the boundary, that is $J_n'(\beta a) = 0$. This restricts β to the values ξ_{nm}/a, where ξ_{nm} is the mth zero of $J_n'(\xi)$. Equation (8.40) may then be written $c^{-2}\omega^2 = a^{-2}\xi_{nm}^2 + \gamma^2$. It is the dispersion relation for the nmth mode of propagation of waves of wave-length $2\pi/\gamma$ along a circular wave guide of radius a.

We can also find the normal modes of vibration inside a cylinder of length L with closed ends by writing $\gamma = N\pi/L$, where N is an integer, as in (8.22).

8.4.2 Radiating cylindrical waves

Consider the solution of (8.39) in the region *outside* the infinite cylinder $r = a$. The function Y_n is now included in the solution, because its singularity at $r = 0$ is not within the region. A **radiating** wave is a solution which behaves asymptotically as $r \to \infty$ like an outgoing wave. Consider the function

$$\psi(r, t) = [J_n(\beta r) + iY_n(\beta r)]\,e^{-i\omega t} .$$

From the formulae (13.19) we see that the asymptotic behaviour of ψ is

$$\psi \sim \left(\frac{2}{\pi\beta r}\right)^{\frac{1}{2}} e^{-i(2n+1)\pi/4}\, e^{i(\beta r-\omega t)} \quad .$$

This is an outgoing wave radiating with speed $\beta^{-1}\omega$ and attenuating as $r^{-\frac{1}{2}}$. Similarly the function $(J_n - i\,Y_n)e^{-i\omega t}$ represents an incoming wave at large distances. The solution (8.40), which is explicitly a linear superposition of *stationary* waves, could clearly be rearranged as a linear superposition of the radiating and incoming waves we have here defined.

8.4.3 Radiation from a cylindrical source

Suppose the boundary $r = a$ is executing some *small* prescribed motion, so that the radial component of its velocity is a known function $U(\theta, z, t)$, and we wish to find the resulting radiation. We first perform a Fourier analysis of the function U with respect to its three arguments, which expresses U as a combination of a summation over n, and possibly integrals over γ and ω. We assume the required solution $\phi(r, \theta, z, t)$ to be a linear superposition of terms like (8.40), with $B = iA$. Then we may identify the coefficients in the expression for $[\partial\phi/\partial r]_{r=a}$ with the corresponding coefficients in the expression for U, and hence find ϕ.
We shall not carry through the full process here, but we shall consider two simple examples.

8.4.4 Radially pulsating cylinder

Here $U(\theta, z, t) = K\cos\omega t$ (K constant) .
The Fourier expression consists of one term only. It is independent of θ and z, and we therefore choose $\phi = \text{Re}\psi$, where

$$\psi = C[J_0(kr) + iY_0(kr)]\, e^{-i\omega t} \quad , \tag{8.41}$$

$ck = \omega$ and C is a complex constant to be found from the condition $\partial\psi/\partial r = Ke^{-i\omega t}$ when $r = a$. See problem 8.7.

8.4.5 Vibrating wire

Suppose the cylindrical surface at $r = a$ is executing small lateral oscillations so that the velocity at every point of the surface is $K\cos\omega t$ in the x-direction. The radial component is then $U = K\cos\theta\cos\omega t$. Again we have a single-term Fourier expansion; here with $n = 1$. Therefore we choose

$$\psi = C[J_1(kr) + iY_1(kr)]\, \cos\theta e^{-i\omega t} \tag{8.42}$$

and find C by demanding that $\partial\psi/\partial r = K\cos\theta e^{-i\omega t}$ when $r = a$. See Problem 8.8.

8.5 SPHERICAL WAVES

The wave equation in spherical polar coordinates (r, θ, ϕ) is

$$\frac{1}{r^2}\frac{\partial}{\partial r}\left(r^2\frac{\partial u}{\partial r}\right)+\frac{1}{r^2\sin\theta}\frac{\partial}{\partial\theta}\left(\sin\theta\frac{\partial u}{\partial\theta}\right)+\frac{1}{r^2\sin^2\theta}\frac{\partial^2 u}{\partial\phi^2}=\frac{1}{c^2}\frac{\partial^2 u}{\partial t^2}\ . \tag{8.43}$$

Separated solutions of this equation can be found. They have the form

$$R(r)\,\Theta\,(\theta)\,{}^{\cos}_{\sin}\,m\phi\;{}^{\cos}_{\sin}\,\omega t,$$

where R and Θ satisfy differential equations of the Bessel and Legendre type with solutions well known and tabulated. Suitable choice of these functions provides us, as in section 8.4, with the normal modes of vibration in a spherical cavity, and with the means to calculate the radiation produced by a spherical source executing a prescribed small motion.

We shall confine ourselves to **axi-symmetric** systems, in which the field does not depend on the angle ϕ, and so $m = 0$. We then find that the equation becomes

$$\frac{(r^2R')'}{R}+\frac{(\sin\theta\Theta')'}{\sin\theta\Theta}+\frac{\omega^2}{c^2}r^2=0\quad .$$

The second term must be constant; if we call it $-n(n+1)$ and write $\cos\theta = \mu$ we obtain

$$\frac{d}{d\mu}\left\{(1-\mu^2)\frac{d\Theta}{d\mu}\right\}+n(n+1)\,\Theta=0\quad . \tag{8.44}$$

This is Legendre's equation, and from section 13.3 we see that the solutions will be continuous at $\theta = 0$ and π $(\mu = \pm 1)$, only if n is an integer, and the required functions are $P_n(\cos\theta)$ $(n = 0, 1, 2, \ldots)$.

The radial equation is

$$(r^2R')' + [k^2r^2 - n(n+1)]\,R = 0\quad ,$$

where $ck = \omega$. The substitutions

$$kr = \rho,\ R(r) = \rho^{-1/2}\,Z(\rho)$$

reduce this to **Bessel's equation** of order $n + \frac{1}{2}$,

$$\rho(\rho Z')' + [\rho^2 - (n+\tfrac{1}{2})^2]\,Z = 0\quad . \tag{8.45}$$

Its general solution is $Z = AJ_{n+1/2}(\rho) + BY_{n+1/2}(\rho)$.

Thus the radial function $R(r)$ is a linear combination of the **spherical Bessel and Neumann functions** which are defined as

$$j_n(\rho) = \left(\frac{\pi}{2\rho}\right)^{\frac{1}{2}} J_{n+\frac{1}{2}}(\rho) \text{ and } n_n(\rho) = \left(\frac{\pi}{2\rho}\right)^{\frac{1}{2}} Y_{n+\frac{1}{2}}(\rho) \quad .$$

The function j_n is regular at the origin, but n_n is singular, so for solutions applicable throughout the interior of a sphere we require $B = 0$. As with the cylinder, radiating solutions in the region outside a sphere are represented by functions of the form

$$[j_n(kr) + in_n(kr)]\ P_n(\cos\theta) . e^{-ikct} \quad . \tag{8.46}$$

The spherical Bessel functions can be expressed in terms of trigonometric and rational functions. For example,

$$j_0(\rho) = \rho^{-1} \sin\rho \text{ and } n_0(\rho) = -\rho^{-1} \cos \rho \quad .$$

Hence the velocity potential outside a uniformly radially pulsating sphere will, from (8.46) with $n = 0$, behave for all values of r as $r^{-1}e^{ik(r-ct)}$.

8.5.1 Point sources

The field $\psi = Ar^{-1}e^{ik(r-ct)}$ is called the field of a **simple source** of **strength** A at the origin; A may be complex so as to describe both the amplitude and phase of the source. The total rate of volume flow through an infinitesimal sphere with centre the origin is $-4\pi Ae^{-ikct}$.

Using (8.24) and (8.26) we see that the intensity of the radiated sound, averaged over one period of oscillation, is $\frac{1}{2}\rho_0 ck^2A^2r^{-2}$. The total power radiated is therefore $2\pi\rho_0 ck^2A^2$, independent of r, so that there is no accumulation of energy.

The field $\psi = \mathbf{A} . \nabla(r^{-1}e^{ikr})\ e^{-ikct}$ is called the field of a **dipole source** of strength $\mathbf{A}$ at the origin. It is the field radiated by a laterally oscillating rigid sphere (see Problem 8.12).

The average intensity of the sound radiated in a direction at an angle θ to $\mathbf{A}$ is $\frac{1}{2}\rho_0 ck^4A^2\cos^2\theta r^{-2}$.

8.5.2 Example: Fluid inside an oscillating sphere

A rigid sphere of radius a is executing small linear oscillations of frequency ω. Find the forced oscillations of the gas contained in the sphere.

Solution

Let the velocity of the sphere be $K\cos\omega t$. Then the radial velocity is $K\cos\omega t\cos\theta$. Now $\cos\theta = P_1(\cos\theta)$; hence the motion of the boundary is already analysed into a one-term series of the separated solutions. We may therefore suppose the velocity potential ϕ to be

$$\phi(r, \theta, t) = Aj_1(kr)\cos\theta\cos\omega t, \text{ where } ck = \omega \quad .$$

We require $\phi_r(a, \theta, t) = K\cos\theta\cos\omega t$.

Hence $Akj_1'(ka) = K$, and the result is

$$\phi(r, \theta, t) = \frac{Kj_1(kr)\cos\theta\cos\omega t}{kj_1'(ka)} \quad .$$

Note that this becomes very large if ka is near a zero of the function j_1'. At such frequencies the cavity resonates with the applied oscillation.

8.6 DIFFRACTION

When waves radiating from a source meet an obstacle, the wave pattern becomes modified. If the wavelength is very small compared with the smallest distance which can be observed with the available apparatus, as is often the case in optics, the obstacle produces a shadow with an apparently sharp edge, and the intensity of radiation at any point is determined by geometrical optics, using rays. Geometrical acoustics is less applicable, but sound rays are observable at very high frequencies, and large obstacles such as buildings cast sound shadows. Closer examination reveals that the edge of a shadow is not sharply defined, but breaks up into **fringes**. This phenomenon is known as diffraction. At the other end of the scale obstacles which are small compared with the wavelength disrupt the waves without producing an identifiable shadow. This is also a diffraction problem, but is usually termed **scattering**, and we consider it in section 8.7.

In acoustics the diffraction problem amounts to solving the scalar wave equation subject to certain boundary conditions. Although electromagnetic waves are vector waves, most optical problems not involving the polarisation of the waves can also be treated by scalar theory. We shall therefore relate the results of this section to optical as well as to acoustical applications.

8.6.1 Diffraction at an aperture in a screen

Let $S + A$ be an infinite surface consisting of a screen S with a finite aperture A; we chose the origin in A. The surface separates space into two parts I and T (see Fig. 8.3).

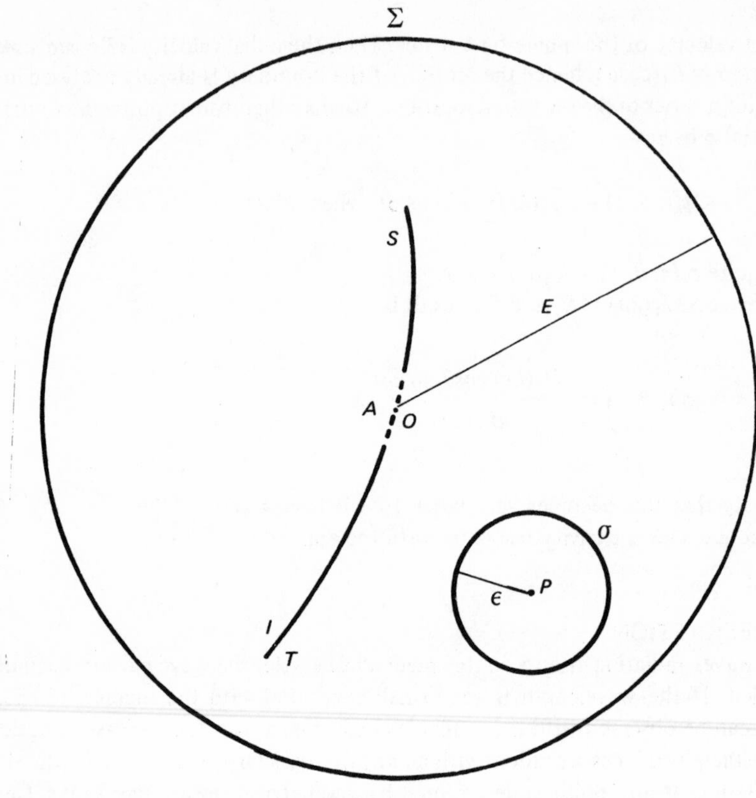

Fig. 8.3

We suppose that in the region I there is a source producing a given incident wave ψ_0, and we wish to determine the radiation transmitted through A into the region T which contains no sources. We consider only monochromatic waves of frequency $\omega = kc$, which have the form $\psi(\mathbf{r})\ e^{-i\omega t}$, and we shall normally omit the factor $e^{-i\omega t}$. We have to solve

$$\nabla^2 \psi + k^2 \psi = 0 \tag{8.47}$$

subject to boundary conditions on S, and fit the solution to the given incident wave. We also recognise that all the radiation in T originates from the region A by imposing the **Sommerfeld radiation conditions**

$$\text{(i) } r\psi \text{ is bounded, (ii) } r\left(\frac{\partial \psi}{\partial r} - ik\psi\right) \to 0, \text{ as } r \to \infty \quad . \tag{8.48}$$

This means that ψ behaves like $r^{-1}e^{ikr}$ at infinity, as if the region A were a point source.

8.6.2 Kirchhoff's method

Let ψ_P be the value of ψ at a point P. We aim to derive a formula for ψ_P in terms of the values of ψ and its normal derivative on $S + A$. Let σ be a small sphere with centre P and radius ϵ, and let Σ be a large sphere with centre O and radius E. Let V be the region, with boundary B, enclosed by $S + A$, σ and Σ. Then, if ϕ is a solution of (8.47),

$$\int_B (\phi\nabla\psi - \psi\nabla\phi).\mathrm{d}\mathbf{S} = \int_V \mathrm{div}(\phi\nabla\psi - \psi\nabla\phi)\mathrm{d}V = \int_V (\phi\nabla^2\psi - \psi\nabla^2\phi)\mathrm{d}V = 0\,. \tag{8.49}$$

Now let Q be any point, and $PQ = R$. Choose the function ϕ so that $\phi_Q = R^{-1}e^{ikR}$. Then the radiation condition (8.48) ensures that

$$\int_\Sigma (\phi\nabla\psi - \psi\nabla\phi).\mathrm{d}\mathbf{S} \to 0 \text{ as } E \to \infty \quad .$$

On the sphere σ, the normal derivative operator in (8.49) is $-\,\partial/\partial R$. Only terms of order R^{-2} will yield a non-zero integral as $\epsilon \to 0$. Also, as $\epsilon \to 0$, $\psi \to \psi_P$. Hence the contribution from σ to the integral is $-4\pi\psi_P$.
Then (8.49) gives

$$4\pi\,\psi_P = \int_{S+A} \left(\psi\frac{\partial}{\partial n}\frac{e^{ikR}}{R} - \frac{\partial\psi}{\partial n}\frac{e^{ikR}}{R}\right)\mathrm{d}S \quad , \tag{8.50}$$

where $\partial/\partial n$ denotes the normal derivative from I *into* T.

We could find ψ at all points in T if we knew the values of ψ and $\partial\psi/\partial n$ on $S + A$, but these values are not provided by the physical boundary conditions. Indeed, if they were, the mathematical problem would be improperly formulated, for (8.47), subject to (8.48), does not permit the arbitrary prescription of both ψ and $\partial\psi/\partial n$ on $S + A$.

Kirchhoff's approximation is to assume that

$$\begin{aligned} &\psi = 0,\ \partial\psi/\partial n = 0 \text{ on } S \quad , \\ &\psi = \psi_0,\ \partial\psi/\partial n = \partial\psi_0/\partial n \text{ on } A \quad . \end{aligned} \tag{8.51}$$

This assumption is inconsistent, and the function ψ thus computed, using (8.50), fails in general to satisfy the boundary conditions (8.51). However, the approximation is known to work when the aperture is large compared with the wavelength and ψ_P is not required when P is near the screen, which is usually the case in optical diffraction.

Let $\tilde{\psi}_P$ be the function computed by (8.50) with S and A interchanged in (8.51). Then we see that $\psi_P + \tilde{\psi}_P$ is the function obtained in the absence of the screen, that is, $\psi + \tilde{\psi} = \psi_0$. This is **Babinet's principle of complementary screens**: a thin opaque obstacle produces a diffracted wave identical to the wave emitted from an aperture of the same form. This wave **interferes** with the incident wave, giving $\psi_0 - \psi$ in T, from which the shadow pattern is derived. Babinet's principle applies only when the Kirchhoff approximation is appropriate.

8.6.3 Rayleigh's method. Rigid screen

The aforementioned inconsistency can be avoided by taking ϕ as the **Green's function** appropriate to the boundary conditions of the physical problem. Taking the case of acoustic waves and a rigid screen we define the function $G(P, Q)$ of the coordinates of the points P and Q as follows:

$$\nabla_Q^2 G + k^2 G = 0 \text{ in } T \quad ,$$

$$\frac{\partial_Q G}{\partial n} = 0 \text{ on } S + A \quad ,$$

$$G \to \frac{e^{ikR}}{R} \text{ as } Q \to P \quad ,$$

$$r_Q \left(\frac{\partial G}{\partial r_Q} - ikG \right) \to 0 \text{ as } r_Q \to \infty \quad .$$

Then, using the argument leading to (8.50), we obtain

$$4\pi \psi_P = \int_{S+A} \left(\psi \frac{\partial G}{\partial n} - \frac{\partial \psi}{\partial n} G \right) \mathrm{d}S$$

$$= - \int_{S+A} \frac{\partial \psi}{\partial n} G \, \mathrm{d}S \quad .$$

The boundary conditions applicable to the rigid screen are:

On S: $\quad \psi$ not specified, $\dfrac{\partial \psi}{\partial n} = 0 \quad .$

On A: $\quad \psi = \psi_0, \; \dfrac{\partial \psi}{\partial n}$ not specified.

Hence $$4\pi\psi_P = - \int_A \frac{\partial \psi}{\partial n} G \mathrm{d}S \quad . \tag{8.52}$$

Like (8.50), this is not an explicit solution since we have no knowledge of $\partial\psi/\partial n$ on A. However, if P is on A, $\psi_P = \psi_0(P)$, which is given, and so (8.52) becomes an **integral equation** for the function $\partial\psi/\partial n$ on A which can in principle be solved. The whole procedure is consistent and the result will satisfy the correct boundary conditions. Its disadvantage is that the Green's function G can rarely be found, except for plane screens; for most other cases either Kirchhoff's approximation or much more complicated integral equation techniques must be applied.

8.6.4 Plane rigid screen

Let the screen be in the plane $z = 0$, and let the wave $\psi_0(x, y, z)$ be incident from the left ($z < 0$). If there is no aperture the solution for which the normal derivative vanishes on S is easily seen to be

$$\begin{Bmatrix} \psi_0(x, y, z) + \psi_0(x, y, -z) & (z < 0) \\ 0 & (z > 0) \end{Bmatrix}.$$

The solution $\phi(x, y, z)$ to the problem of diffraction at an aperture A is then

$$\begin{aligned} \phi(x, y, z) &= \psi_0(x, y, z) + \psi_0(x, y, -z) - \psi(x, y, -z) \quad (z < 0) \\ \phi(x, y, z) &= \psi(x, y, z) \quad (z > 0). \end{aligned} \tag{8.53}$$

Here ψ is to be found from (8.52). It satisfies (i) $\nabla^2\psi + k^2\psi = 0$ (ii) the radiation conditions (8.48), (iii) $\psi_z(x, y, +0) = 0$ on S, (iv) $\psi(x, y, 0) = \psi_0(x, y, 0)$ on A. The three terms in (8.53) are the incident, reflected, and diffracted waves in the region I. It can be verified that the form of the last term is dictated by the conditions on S and at infinity and the requirement that $\phi(x, y, z)$ and $\phi_z(x, y, z)$ are continuous on A.

By symmetry we see that the Green's function required is

$$G = \frac{e^{ikR}}{R} + \frac{e^{ikR_m}}{R_m},$$

where $R = PQ$, $R_m = P_m Q$, and P_m is the mirror image of P in the plane $z = 0$.

Then, on A, $G = \dfrac{2e^{ikR}}{R}$, and hence

$$2\pi\psi_P = -\int_A \frac{\partial\psi}{\partial z}\frac{e^{ikR}}{R}\, dS . \tag{8.54}$$

Explicitly, writing $f(x, y)$ for the unknown function $\psi_z(x, y, 0)$, we have the prescription:

Find $f(x, y)$, the solution of the integral equation

$$2\pi\psi_0(x, y, 0) = -\iint_A f(\xi, \eta)\,\frac{e^{ik\sqrt{[(\xi-x)^2+(\eta-y)^2]}}}{\sqrt{[(\xi-x)^2+(\eta-y)^2]}}\,d\xi\,d\eta \quad . \tag{8.55}$$

Then, for $z > 0$,

$$2\pi\psi(x, y, z) = -\iint_A f(\xi, \eta)\,\frac{e^{ik\sqrt{[(\xi-x)^2+(\eta-y)^2+z^2]}}}{\sqrt{[(\xi-x)^2+(\eta-y)^2+z^2]}}\,d\xi\,d\eta \quad . \tag{8.56}$$

For an aperture large compared with the wavelength it suffices to approximate $\partial\psi/\partial z$ in (8.54) by the incident-wave values, giving **Rayleigh's approximation**

$$2\pi\psi_P = -\int_A \frac{\partial\psi_0}{\partial z}\,\frac{e^{ikR}}{R}\,dS \quad . \tag{8.57}$$

This gives results close to Kirchhoff's approximation, but has the advantages of simplicity and mathematical consistency.

Equation (8.57) may be regarded as an approximate application, and (8.54) as an exact application, of **Huyghens' principle** to the problem. Huyghens' principle states that each point of an advancing wave front acts as a point source. We see (8.57) as an expression of ψ_P as a linear combination of waves from simple sources on A of strength $-\dfrac{1}{2\pi}\dfrac{\partial\psi_0}{\partial z}\,dS$. A similar interpretation may be placed on (8.50).

Another form of Rayleigh's approximation is obtained if we take the boundary condition $\phi = 0$ on S, which is preferred in optics (Sommerfeld 1954). A discussion of the various approximations is given by Bouwkamp (1954). See also Rayleigh (1896).

8.6.5 Fresnel and Fraunhofer diffraction

Consider a plane wave of unit amplitude incident normally on the screen: $\psi_0 = e^{ikz}$. Using the Rayleigh approximation (8.57) we have, for $z > 0$,

$$2\pi\psi(\mathbf{r}) = -\int_A ik\,\frac{e^{ik|\mathbf{r}-\boldsymbol{\zeta}|}}{|\mathbf{r}-\boldsymbol{\zeta}|}\,d\xi\,d\eta \quad ,$$

where $\boldsymbol{\zeta} \equiv (\xi, \eta)$.

Let the greatest diameter of the aperture be d. When $r >> d$ we can approximate

$$k|\mathbf{r} - \boldsymbol{\zeta}| = kr - \frac{k\,\boldsymbol{\zeta}\,.\,\mathbf{r}}{r} + \frac{k(\boldsymbol{\zeta}\times\mathbf{r})^2}{2r^3} + \text{higher powers of } \zeta/r \quad .$$

The third term is of order kd^2/r. It may be neglected at sufficiently large distances; this is called **Fraunhofer diffraction.** It requires r/d to be much greater than the number of wavelengths spanning the aperture. If kd^2/r is not small we have **Fresnel diffraction,** which also occurs if ψ_0 is the radiation from a point source at a distance not large compared with kd^2.

We shall consider only Fraunhofer diffraction. Neglecting ζ/r in the denominator we obtain the approximation

$$\psi(x, y, z) = -\frac{ike^{ikr}}{2\pi r}\iint_A e^{-ikr^{-1}(\xi x+\eta y)}\mathrm{d}\xi\,\mathrm{d}\eta \quad .$$

8.6.6 Rectangular aperture

If A is the region $z = 0$, $|x| < a$, $|y| < b$ the integral is elementary. We obtain

$$2\pi i\psi = 4ab\,\frac{e^{ikr}}{r}\,\frac{\sin X}{X}\,\frac{\sin Y}{Y} \tag{8.58}$$

where $\quad X = kaxr^{-1}$ and $Y = kbyr^{-1} \quad .$

The intensity is proportional to $|\psi|^2$, and study of the function $(X^{-1}\sin X)^2$ shows that there is a rectangular pattern of intensity maxima, and that nearly all the radiation falls within the rectangle $|X| < \pi$, $|Y| < \pi$. The effective angular width of the transmitted beam is therefore $\lambda/2a$ in the x-direction and $\lambda/2b$ in the y-direction. The method is valid only for large ka and kb, so that all the radiation is at a very narrow angle to the z-axis.

8.6.7 Circular aperture

Let A be the region $z = 0, x^2 + y^2 < a^2$. Let OP be at an angle θ to the z-axis and $\boldsymbol{\zeta}$ at an angle ϕ to the plane of OP and OZ. Then $\boldsymbol{\zeta}\,.\,\mathbf{r} = \zeta r\cos\phi\,\sin\theta$, and we have

$$\psi_P = \frac{-ike^{ikr}}{2\pi r}\int_0^a\int_0^{2\pi} e^{-ik\zeta\sin\theta\,\cos\phi}\zeta\,\mathrm{d}\phi\,\mathrm{d}\zeta \quad .$$

Using (8.61) we find that the angular integral is $2\pi\zeta J_0\,(k\zeta\,\sin\theta)$, and the radial integral is then evaluated with the help of (13.10). The result is

$$\psi_P = -\frac{a^2ke^{ikr}}{r}\,\frac{J_1(ka\sin\theta)}{ka\sin\theta} \quad . \tag{8.59}$$

The intensity $|\psi_P|^2$ has a large maximum at $\theta = 0$, and zeros approximately at the points where $ka\sin\theta = (n+\frac{1}{4})\pi$ $(n = 1, 2, \ldots)$, between which are much smaller and rapidly diminishing maxima. Hence the diffraction pattern consists of a series of rings.

8.6.8 Very small circular aperture

If ka is not large a solution of the integral equation (8.55) must be sought. Taking the case of an aperture much *smaller* than the wavelength ($ka \ll 1$), we approximate by neglecting $ik\ |\mathbf{r} - \boldsymbol{\zeta}|$ in the numerator, so that the integral equation becomes

$$2\pi = -\iint_A \frac{f(\xi, \eta)}{|\mathbf{r} - \boldsymbol{\zeta}|}\, \mathrm{d}\xi\, \mathrm{d}\eta, \text{ for all } \mathbf{r} \text{ in } A.$$

This is equivalent to the problem of finding the electrostatic charge distribution on a charged conducting disc, and the solution is $f(\zeta) = -2\pi^{-1}(a^2 - \zeta^2)^{-1/2}$. Now, taking $R = r$ in (8.54) because the aperture is small, we find $\psi_P = \dfrac{2a}{\pi}\dfrac{e^{ikr}}{r}$.

Radiation is emitted uniformly in all directions.

8.7 SCATTERING

The diffracted waves radiating from an obstacle which is small compared with the incident wavelength exhibit wide angular distribution, and the process is aptly described as scattering. The approximations of diffraction theory are inapplicable, and in general a complicated integral equation must be solved. In a very few cases the Helmholtz equation (8.47) can be solved by separating it in curvilinear coordinates suited to the boundary conditions. We treat here the simplest cases of cylindrical and spherical boundaries.

8.7.1 Scattering by a cylinder

Let the cylindrical boundary be $x^2 + y^2 = a^2$ and the incident wave $\psi_0 = e^{ikx}$. We first express ψ_0 in terms of the separated solutions of the wave equation in cylindrical coordinates (r, θ, z), where $x = r\cos\theta, y = r\sin\theta$. Now ψ_0 is independent of z and continuous at the origin, and $\nabla^2\psi_0 + k^2\psi_0 = 0$. We may therefore write $\psi_0 = e^{ikr\cos\theta} = \sum_{n=-\infty}^{\infty} A_n J_n(kr)\, e^{in\theta}$.

This is a Fourier series for ψ_0, and so the coefficients are

$$A_n J_n(\rho) = \frac{1}{2\pi}\int_0^{2\pi} e^{i(\rho\cos\theta - n\theta)}\mathrm{d}\theta, \text{ where } \rho = kr \quad .$$

Differentiate n times with respect to ρ and set $\rho = 0$. Then

$$A_n J_n^{(n)}(0) = \frac{i^n}{2\pi} \int_0^{2\pi} \cos^n\theta \, e^{-in\theta} \, d\theta \quad .$$

Hence, using the series expansion (13.5), we obtain $A_n = i^n$, giving

$$e^{i\rho\cos\theta} = \sum_{n=-\infty}^{\infty} i^n J_n(\rho) e^{in\theta} \tag{8.60}$$

and

$$J_n(\rho) = \frac{(-i)^n}{2\pi} \int_0^{2\pi} e^{i(\rho\cos\theta - n\theta)} d\theta \quad . \tag{8.61}$$

The scattering problem is to find a solution of $\nabla^2\psi + k^2\psi = 0$ $(r > a)$ satisfying prescribed boundary conditions on $r = a$ and having the form $\psi = \psi_0 + \psi_s$, where the scattered wave ψ_s must be a radiating wave (section 8.4.2), which may be written

$$\psi_s = \sum_{n=-\infty}^{\infty} i^n c_n \left[J_n(kr) + iY_n(kr)\right] e^{in\theta} \quad , \tag{8.62}$$

and the coefficients c_n are to be found. We take the case of a **rigid boundary**. The condition is $\partial\psi/\partial r = 0$ when $r = a$, for all θ. Clearly each term of the series for $\psi = \psi_0 + \psi_s$ must satisfy a corresponding condition. Hence

$$J_n'(ka) + c_n \left[J_n'(ka) + iY_n'(ka)\right] = 0 \; (n = 0, 1, 2, \ldots) \quad . \tag{8.63}$$

Thus the c_n can be found and the solution expressed as an infinite series.

Using (13.16) and (13.17) we obtain the asymptotic formula

$$\psi_s \sim \left(\frac{2}{\pi k r}\right)^{1/2} e^{i(kr - \pi/4)} \sum_{n=-\infty}^{\infty} c_n e^{in\theta} \qquad (r \to \infty) \quad .$$

The mean intensity of the scattered wave is

$$I_s = \tfrac{1}{2}(-\rho_0 \psi_t)\bar{\psi}_r = \frac{\rho_0 \omega}{\pi r} \left| \sum_{n=-\infty}^{\infty} c_n e^{in\theta} \right|^2 \quad . \tag{8.64}$$

We define the **differential scattering cross section** $\sigma(\theta)$ by saying that $\sigma(\theta)d\theta$ is the power per unit length of the cylinder scattered in directions between θ and $\theta + d\theta$, for unit incident wave intensity.

So
$$\sigma(\theta)=\frac{I_s r}{\frac{1}{2}\rho_0\omega k}=\frac{2}{\pi k}\left|\sum_{n=-\infty}^{\infty} c_n e^{in\theta}\right|^2 . \tag{8.65}$$

In the **long wavelength limit**, $ka << 1$. Examination of (8.63) and the formulae (13.12-14) reveals that neglecting all terms for which $|n| > 1$ creates a relative error of order $(ka)^2$, and we obtain

$$\sigma(\theta)\cong\frac{2}{\pi k}|c_0+c_1e^{i\theta}+c_{-1}e^{-i\theta}|^2\cong\frac{\pi(ka)^3}{8}a(1-2\cos\theta)^2 . \tag{8.66}$$

The scattering is thus mostly backward, with a small peak in the forward direction, and zero radiation at 60°. The **total scattering cross section** σ is the total power per unit length scattered from an incident wave of unit intensity.

$$\sigma=\int_0^{2\pi}\sigma(\theta)\mathrm{d}\theta\cong\tfrac{3}{4}\pi^2(ka)^3a. \qquad (ka << 1) .$$

Since ka is small, this is much less than the geometrical cross section $2a$.

8.7.2 Scattering by a rigid sphere

Using spherical polar coordinates (r, θ, ϕ), where $x = r\sin\theta\cos\phi$, $y = r\sin\theta\sin\phi$, $z = r\cos\theta$, we first express the incident wave, which we take as $\psi_0 = e^{ikz} = e^{ikr\cos\theta}$, in terms of the axisymmetric (independent of ϕ) separated solutions (8.46). Let $\psi_0 = e^{ikr\cos\theta} = \sum_{n=0}^{\infty} A_n j_n(kr) P_n(\cos\theta)$ (n_n does not appear because ψ_0 is continuous at the origin). Writing $kr = \rho$ and $\cos\theta = \mu$, this is seen as an expansion of $e^{i\rho\mu}$ in Legendre polynomials. The coefficients are, from (13.30),

$$A_n j_n(\rho)=(n+\tfrac{1}{2})\int_{-1}^{1}e^{i\rho\mu}P_n(\mu)\,\mathrm{d}\mu .$$

Differentiate n times with respect to ρ and set $\rho = 0$.

$$A_n j_n^{(n)}(0)=(n+\tfrac{1}{2})i^n\int_{-1}^{1}\mu^n P_n(\mu)\,\mathrm{d}\mu .$$

Now $j_n(\rho) = (\pi/2\rho)^{1/2} J_{n+1/2}(\rho)$. Hence, with the aid of the identity $\Gamma(\frac{1}{2}) = \sqrt{\pi}$ and the relations (13.25) and (13.31), from which we obtain the first term in the expansion of μ^n in the Legendre polynomials, we find $A_n = i^n(2n + 1)$.

The solution ψ which we seek therefore has the form

$$\psi=\psi_0+\psi_s=\sum_{n=0}^{\infty}(2n+1)i^nP_n(\cos\theta)\{j_n(kr)+c_n[j_n(kr)+in_n(kr)]\} . \tag{8.67}$$

The scattered wave ψ_s is a radiating wave. The coefficients c_n are found from the condition that $\partial\psi/\partial r$ is zero on $r = a$. We obtain

$$j_n'(ka) + c_n\,[j_n'(ka) + in_n'(ka)] = 0, \quad (n = 0, 1, 2, \ldots) \quad ,$$

from which the c_n can be found and the complete solution (8.67) computed.

Following the argument leading to (8.64) we obtain the intensity I_s of the scattered wave at great distance. We now define the differential scattering cross section $\sigma(\theta, \phi)$ by taking $\sigma(\theta, \phi)\mathrm{d}\Omega$ as the power scattered into a solid angle element $\mathrm{d}\Omega$ at (θ, ϕ) for unit incident wave intensity. Clearly, by the symmetry of the problem, $\sigma(\theta, \phi)$ does not here depend on ϕ. Write $\sigma(\theta, \phi) = \sigma(\theta)$. Then

$$\sigma(\theta) = \frac{I_s\, r^2}{\frac{1}{2}\rho_0 \omega k} = \frac{1}{k^2}\left|\sum_{n=0}^{\infty} (2n+1)\, c_n P_n(\cos\theta)\right|^2 \quad .$$

In the long wavelength limit the first approximation is obtained using only the first two terms of the series, and the formula corresponding to (8.66) is $\sigma(\theta) \cong \frac{1}{9}\,(ka)^4 a^2 (1 - 3\,\cos\theta)^2$. The total scattering cross section σ is obtained by integrating over all solid angles:

$$\sigma = \int \sigma \mathrm{d}\Omega = \int_0^{\pi} 2\pi\sin\theta \quad \sigma(\theta)\mathrm{d}\theta = \frac{16\pi}{9}\,(ka)^4 a^2 \quad .$$

Scattering varying as the fourth power of the frequency is called **Rayleigh scattering**; in optics, the blue end of the spectrum is scattered much more strongly than the red. It occurs when light is scattered by molecules or by very short-range fluctuations in gas or liquid density, which act as scattering objects small compared with the wavelength. The blue of the sky and the blue appearance of small-particle smoke are examples of Rayleigh scattering.

8.7.3 Short wave scattering

If ka is not small the series (8.67) converges very slowly. When $ka \sim 10$ the series solution shows the formation of a shadow behind the sphere. As ka increases further the solution must approach the geometrical acoustics limit, but it becomes complicated by rapidly fluctuating terms which must be smoothed out if the results are to be compared with diffraction theory. For large ka the method used in optics is to divide the sphere into an illuminated region S_{il} and a shaded region S_{sh}, and then to apply some variation of the Kirchhoff approximation so as to obtain the scattered wave as a superposition $\psi_s = \psi_{\mathrm{il}} + \psi_{\mathrm{sh}}$ from Huyghens sources in the two regions. It turns out that, for $ka >> 1$, the reflected wave ψ_{il} is approximately isotropic, while the wave ψ_{sh} is very similar to the wave transmitted through a circular aperture.

PROBLEMS

8.1 In a fluid at rest under gravity a plane wave propagates, the velocity potential being $\phi = A\cos(\mathbf{k}\,.\,\mathbf{r} - \omega t)$, where $\omega = c|\mathbf{k}|$ and c is the velocity of sound. Using (8.10) and (8.13), show that the conditions for linearisation (8.16) are satisfied if (i) the local speed $|\mathbf{v}|$ is small compared with c and also (ii) both the wavelength and the vertical extent of the domain under consideration are small compared with c^2/g.

8.2 Show that harmonic sound waves travelling along a rectangular tube with rigid walls have the form

$$\phi_{mn} = \cos\frac{m\pi x}{a}\cos\frac{n\pi y}{b}\,e^{i(kz-\omega t)} \quad ,$$

where $$\omega^2 = c^2k^2 + \pi^2\left(\frac{m^2}{a^2} + \frac{n^2}{b^2}\right)$$

and $m, n = 0, 1, 2, \ldots$.

8.3 Two narrow tubes T_1, T_2 of cross section S_1, S_2, containing immiscible fluids of density ρ_1, ρ_2 in which the velocities of sound are c_1, c_2, are joined at O. Show that the apportionment of energy in waves reflected and transmitted at O is as in section 8.3.3, but with S replaced by Z^{-1} throughout where $Z = \rho c/S$ is the characteristic impedance.

8.4 A fluid is contained in the space between two rigid concentric cylinders, radii a and b, and two plane ends normal to the axis of the cylinders. It vibrates in a mode in which the velocity is everywhere perpendicular to the axis and radially directed. Show that the normal frequencies are the values of kc, where k is a solution of the equation

$$J_1(ka)\,Y_1(kb) = J_1(kb)\,Y_1(ka) \quad .$$

8.5 Verify that $r^{-1}e^{ikr}$ is a solution of $(r^2R')' + k^2r^2R = 0$. Show that the frequencies of the spherically symmetric normal modes of a gas confined within a fixed rigid sphere of radius a are given by $\omega = kc$, where k satisfies $\tan ka = ka$.

8.6 Fluid is confined between concentric spherical boundaries. The outer sphere $r = b$ is fixed. The inner sphere pulsates so that its radius is $a + \epsilon\cos kct$, where ϵ is small. Show that the velocity potential of the forced motion is

$$\phi = \frac{\epsilon a^2 kc}{r}\,\frac{\sin(kb - kr) - kb\cos(kb - kr)}{(1 + abk^2)\sin(kb - ka) - (kb - ka)\cos(kb - ka)}\sin kct \quad .$$

8.7 A long cylinder of mean radius a is pulsating in air, so that the radial velocity at the surface is $K\cos\omega t$ (see section 8.4.4). Assuming that the amplitude of the pulsation is small, show that the radiated velocity potential field is

$$\phi = -\frac{K}{k}\,\frac{[J_1(ka)\,J_0(kr) + Y_1(ka)Y_0(kr)]\cos\omega t + [J_1(ka)Y_0(kr) - Y_1(ka)J_0(kr)]\sin\omega t}{[J_1(ka)]^2 + [Y_1(ka)]^2}$$

8.8 A rigid cylinder is executing small lateral oscillations with velocity $K\cos\omega t$ (see section 8.4.5). Show that the radiated velocity potential field is $\phi = \mathrm{Re}\psi$, where

$$\psi = Ce^{-i\gamma}\cos\theta\,[J_1(kr) + i\,Y_1(kr)]\,e^{-i\omega t} \quad ,$$

$$C = Kk^{-1}\{[J_1'(ka)]^2 + [Y_1'(ka)]^2\}^{-1/2}$$

and $$\gamma = \arg[J_1'(ka) + i\,Y_1'(ka)] \quad .$$

8.9 A fluid of density ρ is confined to a rectangular tube with three rigid sides $z = 0$, $z = b$, $y = -h$ and one elastic side which in equilibrium is $y = 0$ but can move in the y-direction. The fluid is subject to no external forces other than those at the tube boundary. When the displacement of the elastic side is small and of the form $y = \eta(x, t)$, the pressure $P(x, y, z, t)$ may be taken to satisfy the boundary condition $P(x, 0, z, t) = \rho K\,\eta(x, t)$ where K is constant. Show that the dispersion relation for velocity potential waves of the form

$$\phi(x, y, z, t) = Y(y)\cos(kx - \omega t)$$

is obtained by eliminating q from the equations

$$\omega^2 = c^2(k^2 + q^2) = -Kq\tanh qh \quad .$$

Show graphically that, as K falls from the rigid-boundary value ∞ (elastic wall becoming weaker), the values of q for a given k fall from $N\pi/h$ towards $(N-\frac{1}{2})\pi/h$ ($N = 1, 2, \ldots$), thus depressing the cut-off frequency qc for each mode of propagation. Putting $q = ip$, show that the plane-wave mode is replaced by a mode with $Y \propto \cosh p(y + h)$ and dispersion relation given by $\omega^2 = c^2(k^2 - p^2) = Kp\tanh ph$, in which all frequencies can propagate.
[For a weak boundary, K small, this approaches the incompressible approximation of section 10.3.1].

8.10 Solving (8.34) in the case when $S = Ax^{\alpha}$, where A and α are positive constants, show that when the tube is closed at $x = L$ the normal longitudinal oscillations of the velocity potential are of the form $x^{-m}J_m(kx)\,e^{-ikct}$, where $m = (\alpha - 1)/2$ and k satisfies $J_{m+1}(kL) = 0$.

8.11 A solid sphere of radius a immersed in a fluid is pulsating radially so that the radial velocity of its surface is $K\cos\omega t$, where $K \ll \omega a$. Show that the velocity potential outside the sphere is Reψ, where ψ is the field of a simple source of strength $-Ka^2(1 + k^2a^2)^{-1/2}\,e^{-i\gamma}$, where $\gamma = ka - \tan^{-1}ka$.

8.12 A rigid sphere of radius a is executing small oscillations about the origin along the z-axis with velocity $K\cos\omega t = \text{Re}Ke^{-i\omega t}$. Using (13.11), show that $j_1 = -j_0'$, and hence show that the velocity potential is of the form Reψ where $\psi = A\cos\theta(\mathrm{d}/\mathrm{d}r)(e^{ikr}/r)$. Find the strength of the corresponding dipole source, and show that the radiated intensity at an angle θ to the z-axis is

$$\tfrac{1}{2}\,\rho_0 cK^2a^6k^4(4 + k^4a^4)^{-1}r^{-2}\cos^2\theta \quad .$$

Show that the velocity potential $\phi(r, \theta, t) = L(M\cos\gamma + N\sin\gamma)$, where

$$\begin{aligned} L &= ka^3(4 + k^4a^4)^{-1}r^{-2}\cos\theta \quad , \\ M &= -2 + k^2a^2 - 2k^2ar \quad , \\ N &= k(2a - 2r + k^2a^2r) \quad , \\ \gamma &= k(r - a - ct) \quad . \end{aligned}$$

CHAPTER 9

Electromagnetic Waves

9.1 THE ELECTROMAGNETIC FIELD

The electromagnetic field (**E**, **B**) in a region of space is defined by the behaviour of a charged particle. If the particle carries a charge q and is moving with velocity **v**, it is found to be subject to a force of the form

$$q(\mathbf{E} + \mathbf{v} \times \mathbf{B}) \quad , \tag{9.1}$$

called the Lorentz force. We say that **E** is the electric field and **B** is the magnetic induction field. The electromagnetic field itself is generated by charges and currents. In a material medium, which consists of a large number of charged particles, the fields vary rapidly in the neighbourhood of any of the constituent particles. However, it is possible to define smooth functions to describe macroscopic phenomena, just as in fluid dynamics a velocity field is defined by averaging the rapidly fluctuating individual molecular motions.

9.1.1 Maxwell's equations

We consider a **linear isotropic medium**, in which the effects of the medium on the fields depend only the **permittivity** $\epsilon(\mathbf{r})$ and the **permeability** $\mu(\mathbf{r})$ of the medium. We take **E** and **B** now as the smoothed-out fields and define, for convenience, the magnetic field $\mathbf{H} = \mu^{-1}\mathbf{B}$. Let $\rho(\mathbf{r}, t)$ be the electric charge density (smoothed out), and $\mathbf{j}(\mathbf{r}, t)$ the current density. Then the electric field **E** and the magnetic field **H** are governed by Maxwell's equations

$$\text{div}\, \epsilon\, \mathbf{E} = \rho \tag{9.2}$$

$$\text{div}\, \mu\, \mathbf{H} = 0 \tag{9.3}$$

$$\text{curl}\, \mathbf{E} = -\mu\dot{\mathbf{H}} \tag{9.4}$$

$$\text{curl}\, \mathbf{H} = \mathbf{j} + \epsilon\, \dot{\mathbf{E}} \quad , \tag{9.5}$$

where a dot denotes $\partial/\partial t$.

These equations, which we take as our starting point, are derived in any standard work on electromagnetic theory (for example, Panofsky 1962).

Now let Q be the total charge within a region V bounded by a closed surface S_1, and let J be the total current flowing through a surface Σ bounded by a closed curve Γ. If we apply the divergence theorem to (9.2) and (9.3) and Stokes's theorem to (9.4) and (9.5) we obtain four relations, involving integrals, which are equivalent to Maxwell's equations:

$$\oint_{S_1} \epsilon \mathbf{E} \cdot d\mathbf{S} = Q \tag{9.6}$$

$$\oint_{S_1} \mu \mathbf{H} \cdot d\mathbf{S} = 0 \tag{9.7}$$

$$\oint_{\Gamma} \mathbf{E} \cdot d\mathbf{r} = -\int_{\Sigma} \mu \dot{\mathbf{H}} \cdot d\mathbf{S} \tag{9.8}$$

$$\oint_{\Gamma} \mathbf{H} \cdot d\mathbf{r} = J + \int_{\Sigma} \epsilon \dot{\mathbf{E}} \cdot d\mathbf{S} \quad . \tag{9.9}$$

Equation (9.6) is Gauss's flux theorem for the electric field, which arises from Coulomb's inverse-square law of force between two static charges. Equation (9.7) asserts the non-existence of free magnetic poles. Equation (9.8) is Faraday's law giving the electromotive force induced in a circuit by a changing magnetic induction field. Equation (9.9) is Ampère's circuital law modified to include the term involving $\epsilon\dot{\mathbf{E}}$ which was introduced by Maxwell and is justified by the behaviour of electromagnetic waves.

9.1.2 Boundary conditions

At an interface between two media the functions $\epsilon(\mathbf{r})$ and $\mu(\mathbf{r})$ are discontinuous and Maxwell's equations cease to apply. However, the integral forms remain valid, and the choice of suitable limiting forms for the surface S_1 and the curve Γ in (9.6) to (9.9) leads to the following results.

Let $\mathbf{n}$ be the unit normal drawn at the interface from region 1 with properties (ϵ_1, μ_1) to region 2 with properties (ϵ_2, μ_2). Suppose that in the interface there is a surface charge density τ and a surface current density $\mathbf{K}$. Then

$$[\epsilon \mathbf{E} \cdot \mathbf{n}]_1^2 = \tau \tag{9.10}$$

$$[\mu \mathbf{H} \cdot \mathbf{n}]_1^2 = 0 \tag{9.11}$$

$$[\mathbf{n} \times \mathbf{E}]_1^2 = 0 \tag{9.12}$$

$$[\mathbf{n} \times \mathbf{H}]_1^2 = \mathbf{K} \quad . \tag{9.13}$$

9.1.3 Energy flow in the field

The flow of energy in a varying electromagnetic field is described by the **Poynting vector**

$$\mathbf{N} = \mathbf{E} \times \mathbf{H} \quad . \tag{9.14}$$

We can see this by expanding div **N** and making use of Maxwell's equations.

$$\text{div}\,(\mathbf{E} \times \mathbf{H}) = \mathbf{H} \,.\, \text{curl}\, \mathbf{E} - \mathbf{E} \,.\, \text{curl}\, \mathbf{H}$$

$$= \mathbf{H} \,.\, (-\mu \dot{\mathbf{H}}) - \mathbf{E} \,.\, (\mathbf{j} + \epsilon \dot{\mathbf{E}}) \quad .$$

Hence, using the well-known formula for the electromagnetic energy density

$$U = \tfrac{1}{2}(\epsilon\, \mathbf{E}^2 + \mu\, \mathbf{H}^2) \quad , \tag{9.15}$$

and noting that the rate of heat dissipation per unit volume, i.e. the **Joule loss**, is **j**. **E**, we arrive at the local conservation law

$$\dot{U} + \mathbf{j} \,.\, \mathbf{E} + \text{div}\, \mathbf{N} = 0 \quad . \tag{9.16}$$

Integrating over a region V bounded by a closed surface S, we have

$$\frac{\mathrm{d}}{\mathrm{d}t} \int_V \frac{\epsilon\, \mathbf{E}^2 + \mu \mathbf{H}^2}{2}\, \mathrm{d}V + \int_V \mathbf{j} \,.\, \mathbf{E}\, \mathrm{d}V + \oint_S \mathbf{E} \times \mathbf{H} \,.\, \mathrm{d}\mathbf{S} = 0 \quad . \tag{9.17}$$

This asserts the conservation of energy. It states that the rate of increase of electromagnetic energy in V, together with the rate of Joule loss in V, is equal to $-\oint_S \mathbf{N} \,.\, \mathrm{d}\mathbf{S}$. But this quantity is the total flux of the field **N** *entering* V. Hence the Poynting vector **N** may be regarded as the energy flux intensity. This notion is only valid in the context of an integration of **N** over a closed surface, and so caution is required in its interpretation. It is, for example, nonsensical to associate a flow of energy with crossed static electric and magnetic fields. We shall henceforward assume the validity of the interpretation in the case of travelling waves.

9.2 THE WAVE EQUATION

In a uniform medium, in which ϵ and μ are constant, the fields **E** and **H** satisfy the inhomogeneous classical wave equation. Taking the curl of (9.4) and using (9.5), we obtain

$$\nabla^2 \mathbf{E} - \mu\epsilon\, \ddot{\mathbf{E}} = \epsilon^{-1}\, \nabla\rho + \mu\, \partial\mathbf{j}/\partial t \quad . \tag{9.18}$$

The solution will in general consist of a free wave solution of the homogeneous wave equation together with a forced term due to the source fields ρ and **j**. Not all solutions of (9.18) represent possible electric fields, for **E** must also satisfy (9.2); this extra condition also ensures the existence of a corresponding magnetic field **H** such that the remaining Maxwell's equations can be satisfied.

9.2.1 Conducting medium

In the case of a **uniform linear isotropic conductor** the current field **j** is not arbitrary, but satisfies Ohm's law

$$\mathbf{j} = \sigma \mathbf{E} \quad , \tag{9.19}$$

where σ is the conductivity. In the absence of accumulation of charge within the medium, (9.18) becomes

$$\nabla^2 \mathbf{E} = \mu\epsilon\, \ddot{\mathbf{E}} + \mu\sigma\dot{\mathbf{E}} \quad . \tag{9.20}$$

In an **insulator,** σ/ϵ is small and we have the classical wave equation, with wave velocity $c = (\mu\epsilon)^{-1/2}$.

In a **good conductor,** σ/ϵ is large and we have the diffusion equation, an equation of the parabolic type (see section 11.4) which does not possess travelling wave solutions.

Notice that ϵ/σ has the dimension of time. It is called the **relaxation time.** The validity of the above approximations therefore depends on the rapidity of changes in **E**; in particular we see that a medium which is a good conductor at low frequencies can become a bad conductor at high frequencies.

When both terms of (9.20) are significant, the equation is hyperbolic and there are travelling wave solutions, but they will be damped or attenuated.

In a **perfect conductor** ($\sigma = \infty$), equation (9.19) can be satisfied only if $\mathbf{E} = 0$ identically. Maxwell's equations then reduce to the magnetostatic equations

$$\dot{\mathbf{H}} = 0, \ \text{div}\, \mathbf{H} = 0, \ \text{curl}\, \mathbf{H} = \mathbf{j} \tag{9.21}$$

9.2.2 Plane Waves

Let **E** depend only on t and the coordinate x. Then, in an insulator containing no accumulations of charge,

$$\frac{\partial^2 \mathbf{E}}{\partial x^2} = \frac{1}{c^2}\,\frac{\partial^2 \mathbf{E}}{\partial t^2}, \ \text{where } c^2 = \frac{1}{\mu\epsilon} \quad .$$

Also $$\frac{\partial \mathbf{E}\,.\,\mathbf{i}}{\partial x} = \text{div}\, \mathbf{E} = 0 \quad ,$$

where $\mathbf{i}$ is the unit vector in the x-direction. The general solution is a plane wave propagating in the x-direction:

$$\mathbf{E} = \mathbf{f}(x - ct) + \mathbf{g}(x + ct) \quad ,$$

where $\mathbf{f}$ and $\mathbf{g}$ are arbitrary vector functions satisfying $\mathbf{f}'.\mathbf{i} = \mathbf{g}'.\mathbf{i} = 0$. Therefore, except for an uninteresting uniform component, the solution consists of the superposition of two **transverse** vector waves travelling in opposite directions with speed c.

For the single wave

$$\mathbf{E} = \mathbf{f}(x - ct) \tag{9.22}$$

we have $\quad -\mu\dot{\mathbf{H}} = \text{curl}\,\mathbf{E} = \mathbf{i} \times \mathbf{f}'(x - ct)$

giving $\quad \mathbf{H} = \epsilon c\,\mathbf{i} \times \mathbf{f}(x - ct) \quad .$

Hence the magnetic field is both transverse and perpendicular to $\mathbf{E}$, and $\mathbf{i}, \mathbf{E}, \mathbf{H}$ form a right-handed triad.

$$\text{The energy density } U = \tfrac{1}{2}(\epsilon\,\mathbf{E}^2 + \mu\,\mathbf{H}^2) = \epsilon\mathbf{E}^2 \quad . \tag{9.23}$$

$$\text{The Poynting vector } \mathbf{N} = \mathbf{E} \times \mathbf{H} = c\epsilon\,\mathbf{E}^2\,\mathbf{i} = Uc\,\mathbf{i} \quad , \tag{9.24}$$

which shows that the energy is transported with velocity c.

If in (9.22) the vector function $\mathbf{f}$ is constant in direction the wave is said to be **linearly polarised** (sometimes plane polarised) in that direction. The magnetic field is also constant in direction. The plane of $\mathbf{E}$ and $\mathbf{i}$ is called the plane of polarisation.

9.2.3 Harmonic plane waves

In a harmonic plane wave both transverse components of $\mathbf{f}$ in (9.22) are sinusoidal functions with the same period. An equivalent general expression for a harmonic plane wave proceeding in an arbitrary direction is

$$\mathbf{A}\cos(\mathbf{k}\,.\,\mathbf{r} - \omega t) + \mathbf{B}\sin(\mathbf{k}\,.\,\mathbf{r} - \omega t) \quad , \tag{9.25}$$

where $\omega = c\,|\mathbf{k}|$ and $\mathbf{A}\,.\,\mathbf{k} = \mathbf{B}\,.\,\mathbf{k} = 0$. This wave has frequency ω and it travels in the direction of its **propagation vector k**. Generally it is an **elliptically polarised wave,** for, when $\mathbf{r}$ is fixed, the locus of the end of the vector $\mathbf{E}$ is an ellipse. In the special case when $\mathbf{A}$ and $\mathbf{B}$ are parallel it is a plane polarised wave. If $|\mathbf{A}| = |\mathbf{B}|$ and $\mathbf{A}\,.\,\mathbf{B} = 0$ it is a **circularly polarised** wave.

It is usually more convenient to represent a plane harmonic wave in the form

$$\mathbf{E} = \mathbf{C}\,e^{i(\mathbf{k}.\mathbf{r} - \omega t)} \quad , \tag{9.26}$$

where **C** is a complex vector. Then, taking $\mathbf{C} = \mathbf{A} - i\mathbf{B}$, where **A** and **B** are real the wave (9.25) may be written Re**E** (cf. Problem 1.1).

The **mean energy flow** in a harmonic wave is found by substituting (9.25) into (9.24) and averaging over one period of oscillation. It is

$$\tfrac{1}{2}\epsilon c(\mathbf{A}^2 + \mathbf{B}^2)\,\mathbf{k} = \tfrac{1}{2}\epsilon c\ \mathbf{C}.\bar{\mathbf{C}}\mathbf{k} \quad . \tag{9.27}$$

9.3 REFLECTION AND TRANSMISSION AT A PLANE BOUNDARY

Suppose an insulator 1 occupies the region $x < 0$, and an insulator 2 the region $x > 0$. As the magnetic permeability is close to the free-space value μ_0 for most media, we shall take $\mu = \mu_0$ for both regions. This assumption will simplify the derivations which follow, but their modification for the case of different permeabilities is quite straightforward. Let the permittivities of the two regions be ϵ_1 and ϵ_2, and the speed of light c_1 and c_2. We define the **refractive index** from 1 to 2 as $n_{12} = c_1/c_2$; here we denote it by n. Note that

$$\frac{\epsilon_2 c_2}{\epsilon_1 c_1} = \frac{\mu_0 \epsilon_2 c_2}{\mu_0 \epsilon_1 c_1} = \frac{c_1}{c_2} = n \quad .$$

9.3.1 Normal incidence

Suppose the wave (9.22) with electric vector $\mathbf{f}(x - c_1 t)$ is incident in region 1. To satisfy the boundary conditions we postulate that there is a reflected wave $\mathbf{g}(x + c_1 t)$ in region 1 and a transmitted wave $\mathbf{h}(x - c_2 t)$ in region 2, just as in section 2.5.1. The associated magnetic fields are

$$\epsilon_1 c_1 \mathbf{i} \times \mathbf{f}, \quad -\epsilon_1 c_1 \mathbf{i} \times \mathbf{g} \quad \text{and} \quad \epsilon_2 c_2 \mathbf{i} \times \mathbf{h} \quad .$$

We now apply the boundary conditions (9.12) and (9.13) with $\mathbf{K} = 0$ because no currents flow in the insulators. It is convenient to use these conditions in the form:

The tangential components of **E** and **H** are continuous at the boundary.

We obtain

$$\mathbf{f}(-c_1 t) + \mathbf{g}(c_1 t) = \mathbf{h}(-c_2 t)$$

and

$$\epsilon_1 c_1 \mathbf{f}(-c_1 t) - \epsilon_1 c_1 \mathbf{g}(c_1 t) = \epsilon_2 c_2 \mathbf{h}(-c_2 t) \quad ,$$

from which we deduce that

$$\mathbf{g}(\zeta) = \frac{1-n}{1+n}\,\mathbf{f}(-\zeta) \text{ and } \mathbf{h}(\zeta) = \frac{2}{1+n}\,\mathbf{f}(-\zeta) \quad , \tag{9.28}$$

in analogy with the string problem.

9.3.2 Oblique incidence

Now suppose that the incident, reflected and transmitted waves propagate in the directions of the unit vectors $\boldsymbol{\alpha}$, $\boldsymbol{\beta}$ and $\boldsymbol{\gamma}$ (see Fig. 9.1). The electric fields in the three waves are

$$\mathbf{f}(\boldsymbol{\alpha}\,.\,\mathbf{r}-c_1t),\ \mathbf{g}(-\boldsymbol{\beta}\,.\,\mathbf{r}+c_1t),\ \mathbf{h}(\boldsymbol{\gamma}\,.\,\mathbf{r}-c_2t)\quad .$$

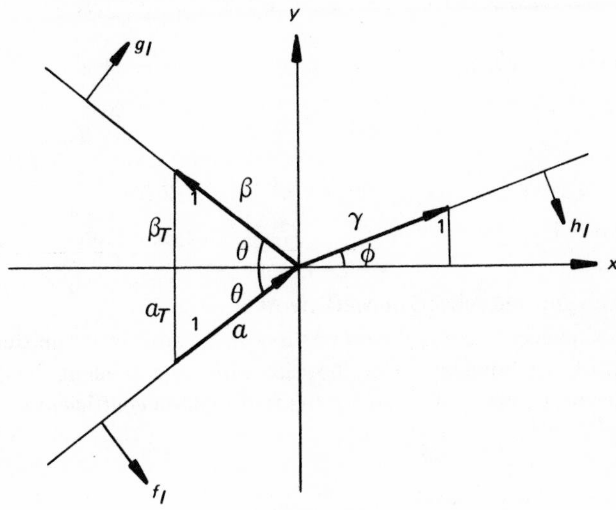

Fig. 9.1

Let $\mathbf{f} = \mathbf{f}^{\perp} + \mathbf{f}^{\mathrm{l}}$, where $\mathbf{f}^{\perp}$ is parallel to the boundary, and similarly decompose $\mathbf{g}$ and $\mathbf{h}$. Let $\boldsymbol{\alpha}_{\mathrm{T}}, \boldsymbol{\beta}_{\mathrm{T}}, \boldsymbol{\gamma}_{\mathrm{T}}$ be the transverse components of $\boldsymbol{\alpha}, \boldsymbol{\beta}, \boldsymbol{\gamma}$, so that $\boldsymbol{\alpha} = \alpha_1\mathbf{i} + \boldsymbol{\alpha}_{\mathrm{T}}$, etc. Then, applying the condition on $\mathbf{E}$ at the boundary $x = 0$, we have

$$\mathbf{f}^{\perp}(\boldsymbol{\alpha}_{\mathrm{T}}\,.\,\mathbf{r}-c_1t)+\mathbf{g}^{\perp}(-\boldsymbol{\beta}_{\mathrm{T}}\,.\,\mathbf{r}+c_1t)=\mathbf{h}^{\perp}(\boldsymbol{\gamma}_{\mathrm{T}}\,.\,\mathbf{r}-c_2t)\quad , \tag{9.29}$$

for all values of t and $\mathbf{r}$. We deduce that

$$\mathbf{f}^{\perp}(-c_1t)+\mathbf{g}^{\perp}(c_1t)=\mathbf{h}^{\perp}(-c_2t) \tag{9.30}$$

and $\quad \boldsymbol{\alpha}_{\mathrm{T}} = \boldsymbol{\beta}_{\mathrm{T}} = n\boldsymbol{\gamma}_{\mathrm{T}}$.

Hence $\boldsymbol{\alpha}$, $\boldsymbol{\beta}$, $\boldsymbol{\gamma}$, $\mathbf{i}$ all lie in the same plane, the **plane of incidence**, $\boldsymbol{\beta}$ makes an angle θ with $-\mathbf{i}$, and $\sin\theta = n \sin\phi$ **(Snell's law of refraction)**.

Taking the directions of $\mathbf{f}^{\mathrm{l}}$, $\mathbf{g}^{\mathrm{l}}$ and $\mathbf{h}^{\mathrm{l}}$ as in Fig. 9.1, and $\mathbf{f}^{\perp}$, $\mathbf{g}^{\perp}$ and $\mathbf{h}^{\perp}$ along the z-axis, we can now express boundary conditions in terms of the corresponding components f^{l}, g^{l}, h^{l}, $f^{\perp}$, $g^{\perp}$, $h^{\perp}$, and (9.30) becomes

$$f^{\perp}(-\zeta)+g^{\perp}(\zeta)=h^{\perp}(-n^{-1}\zeta)\quad .$$

The other condition on **E** gives $f^{\mathrm{I}}\cos\theta - g^{\mathrm{I}}\cos\theta = h^{\mathrm{I}}\cos\phi$. The conditions on **H** give

$$\epsilon_1 c_1 f^{\mathrm{I}} + \epsilon_1 c_1 g^{\mathrm{I}} = \epsilon_2 c_2 h^{\mathrm{I}}$$

$$\epsilon_1 c_1 f^{\perp}\cos\theta - \epsilon_1 c_1 g^{\perp}\cos\theta = \epsilon_2 c_2 h^{\perp}\cos\phi \quad .$$

Solving the four equations for $g^{\perp}$, $h^{\perp}$, g^{I}, h^{I} we obtain the **Fresnel formulae**

$$\frac{g^{\perp}}{f^{\perp}} = \frac{\cos\theta - n\cos\phi}{\cos\theta + n\cos\phi} \qquad \frac{g^{\mathrm{I}}}{f^{\perp}} = \frac{n\cos\theta - \cos\phi}{n\cos\theta + \cos\phi}$$

$$\frac{h^{\perp}}{f^{\perp}} = \frac{2\cos\theta}{\cos\theta + n\cos\phi} \qquad \frac{h^{\mathrm{I}}}{f^{\mathrm{I}}} = \frac{2\cos\theta}{n\cos\theta + \cos\phi} \quad . \tag{9.31}$$

9.3.3 Transmission and reflection coefficients

The transmission coefficient is the proportion of the energy transmitted through a fixed area of the boundary. For the case when the incident electric field is perpendicular to the plane of incidence the transmission coefficient is

$$\tau^{\perp} = \frac{\epsilon_2 c_2 h^{\perp 2}\cos\phi}{\epsilon_1 c_1 f^{\perp 2}\cos\theta} \quad .$$

The **reflection coefficient** $r^{\perp} = g^{\perp 2}/f^{\perp 2}$, and $\tau^{\perp} + r^{\perp} = 1$, as expected by conservation of energy. The quantities τ^{I} and r^{I} are defined similarly.

9.3.4 Total reflection

When $\sin\theta > n$ there is no real vector $\boldsymbol{\gamma}$. No energy is transmitted and the reflected wave is no longer a simple reproduction of the incident wave. The system has to be analysed into its harmonic components. Taking the incident, reflected and transmitted waves as

$$I^{\perp}e^{i(\mathbf{k}.\mathbf{r}-\omega t)}, \quad R^{\perp}e^{i(\mathbf{m}.\mathbf{r}-\omega t)}, \quad T^{\perp}e^{i(\mathbf{l}.\mathbf{r}-\omega t)},$$

we find in this case

$$e^{i\mathbf{l}.\mathbf{r}} = e^{iyk\sin\theta}\, e^{-xk\surd(\sin^2\theta - n^2)} \quad \text{(decaying wave)}$$

$$\frac{R^{\perp}}{I^{\perp}} = \frac{\cos\theta - i\surd(\sin^2\theta - n^2)}{\cos\theta + i\surd(\sin^2\theta - n^2)} \quad .$$

The reflected amplitude is unity, so all the energy is reflected, but the phase is retarded.

9.4 WAVE GUIDES

A wave guide is a hollow metallic tube filled with dielectric. An electromagnetic wave, almost wholly confined to the dielectric, is propagated along the guide. We shall consider an infinite perfectly conducting hollow cylinder S, generated from a closed curve Γ in the plane $z = 0$ by lines parallel to Oz. It contains uniform material (ϵ, μ). We now seek solutions of Maxwell's equations which are of the form

$$\mathbf{E} = \mathbf{F}(x, y)\, e^{i(\beta z - \omega t)}, \mathbf{H} = \mathbf{G}(x, y)\, e^{i(\beta z - \omega t)} \quad . \qquad (9.32)$$

The Maxwell equations curl $\mathbf{E} = i\omega\mu\mathbf{H}$ and curl $\mathbf{H} = - i\omega\epsilon\mathbf{E}$ give

$$i\omega\mu\mathbf{G} = i\beta\hat{\mathbf{z}} \times \mathbf{F} + \text{curl } \mathbf{F} \text{ and } - i\omega\epsilon\mathbf{F} = i\beta\hat{\mathbf{z}} \times \mathbf{G} + \text{curl } \mathbf{G},$$

where $\hat{\mathbf{z}}$ is the unit vector in the z-direction. We shall use a suffix T to denote a component perpendicular to $\hat{\mathbf{z}}$ and write $\mathbf{F}\,.\,\hat{\mathbf{z}} = F_3$ etc. We now take the transverse parts of the these equations, first noting that, because $\mathbf{F}$ does not depend on z, $(\text{curl } \mathbf{F})_{\text{T}} = \nabla F_3 \times \hat{\mathbf{z}}$, and similarly for $\mathbf{G}$.

Hence $\quad i\omega\mu\mathbf{G}_{\text{T}} = i\beta\hat{\mathbf{z}} \times \mathbf{F}_{\text{T}} + \nabla F_3 \times \hat{\mathbf{z}}$

and $\quad -i\omega\mu\mathbf{F}_{\text{T}} = i\beta\hat{\mathbf{z}} \times \mathbf{G}_{\text{T}} + \nabla G_3 \times \hat{\mathbf{z}} \quad ,$

from which we obtain

$$(k^2 - \beta^2)\, \mathbf{F}_{\text{T}} = i\beta\nabla F_3 + i\omega\mu\nabla G_3 \times \hat{\mathbf{z}} \qquad (9.33)$$

and

$$(k^2 - \beta^2)\, \mathbf{G}_{\text{T}} = i\beta\nabla G_3 - i\omega\epsilon\nabla F_3 \times \hat{\mathbf{z}} \quad . \qquad (9.34)$$

Thus we see that the transverse parts of the fields are determined by the two scalar fields F_3 and G_3. If we substitute (9.32) into the Maxwell equations div $\mathbf{E} = 0$ and div $\mathbf{H} = 0$, and use (9.33) and (9.34) we find that

$$\nabla^2 F_3 + (k^2 - \beta^2)\, F_3 = 0 \quad , \qquad (9.35)$$

$$\nabla^2 G_3 + (k^2 - \beta^2)\, G_3 = 0 \quad . \qquad (9.36)$$

These two-dimensional equations are to be solved subject to boundary conditions on Γ. Let $\mathbf{n}$ be the unit normal to Γ. The boundary conditions at the surface of the perfect conductor are that $F_3 = 0$ and $|\mathbf{n} \times \mathbf{F}_{\text{T}}| = 0$, for these are two perpendicular components of the tangential electrical field.

Substituting these conditions into (9.33) gives $\mathbf{n}.\nabla G_3 = 0$. Hence the equations (9.35) and (9.36) are to be solved subject to the boundary conditions

$$F_3 = 0, \mathbf{n}\,.\,\nabla G_3 = 0 \text{ on } \Gamma \qquad (9.37)$$

respectively. We thus have two two-dimensional eigenvalue problems to solve, and the solutions will provide the modes of propagation of waves in the guide. From F_3 and G_3 the complete fields can be found.

Modes in which $F_3 = 0$ are called **transverse electric** (TE) and modes with $G_3 = 0$ are called **transverse magnetic** (TM). The general solution will be a linear combination of all the modes.

We have already considered modes of this type in Chapters 7 and 8. The TM modes correspond to the modes of vibration of a membrane with its periphery fixed, and the TE modes to the modes of propagation of sound waves in tubes. Whereas in the acoustic case there was always a plane-wave solution corresponding to the zero eigenvalue of (9.36), in our present case there is no such solution if Γ is a simple closed curve, for then $\beta^2 = k^2$ gives $F_3 = 0$, and so the transverse electric field $\mathbf{F}_\mathrm{T}$ has to satisfy div $\mathbf{F}_\mathrm{T} = 0$, curl $\mathbf{F}_\mathrm{T} = 0$. Hence we may write $\mathbf{F}_\mathrm{T} = \nabla\psi$, where ψ is a scalar field satisfying $\nabla^2\psi = 0$ and ψ is constant on Γ. The only solution is $\psi =$ constant, i.e. $\mathbf{F}_\mathrm{T} = 0$. However, if the region bounded by Γ is not simply connected (e.g. as in a coaxial cable) the problem has a non-trivial solution, for it is then identical with the problem of determining the electrostatic potential in the neighbourhood of charged conductors. For such modes both F_3 and G_3 are zero, and so they are called TEM modes.

9.4.1 Rectangular wave guide

Let Γ be the boundary of the rectangle $0 \leqslant x \leqslant p$, $0 \leqslant y \leqslant q$. For the TM modes we require $F_3(x, y)$ to satisfy

$$\nabla^2 F_3 + (k^2 - \beta^2) F_3 = 0, F_3 = 0 \text{ on } \Gamma \quad .$$

The solution follows section 7.4.1 and we find the modes

$$F_3 = A \sin \frac{m\pi x}{p} \sin \frac{n\pi y}{q} , \qquad m, n = 1, 2, \ldots$$

with the relation between ω and β, i.e. the dispersion law,

$$\omega^2 = c^2\pi^2 \left(\frac{m^2}{p^2} + \frac{n^2}{q^2}\right) + c^2\beta^2 \quad . \tag{9.38}$$

Thus for each mode there is a lower cut-off frequency below which waves cannot propagate.

When a given frequency is being used, the signal travelling along the guide will be distributed between the various modes of propagation, and suffer differential dispersion and consequent distortion. It is thus advisable to choose the size of the guide so as to admit only the lowest ($p = 1, q = 1$) mode at the given frequency.

Similarly, the TE modes are found to be of the form $G_3 = B\cos(m\pi x/p)$ $\cos(n\pi y/q)$, and to have the same dispersion law (9.38). Thus, for each frequently above the cut-off, there is a TE and a TM mode with the same wavelength. This is a peculiarity of the rectangular guide and is not generally true.

9.4.2 Circular wave guide

We consider the TM modes. Following the treatment in section 7.5 of the circular membrane we use cylindrical coordinates r, θ, z and seek $F_3(r, \theta)$ to satisfy the equation

$$\frac{1}{r}\frac{\partial}{\partial r}\left(r\frac{\partial F_3}{\partial r}\right) + \frac{1}{r^2}\frac{\partial^2 F_3}{\partial \theta^2} + \gamma^2 F_3 = 0 \quad ,$$

(where $\gamma^2 = k^2 - \beta^2$) with the boundary condition on $r = p$, the conducting surface, $F_3(p, \theta) = 0$. The solutions are of the form

$$F_3 = J_m(\gamma r)\,(A\cos m\theta + B\sin m\theta) \quad , \tag{9.39}$$

where $\gamma p = \zeta_{mn}$, the nth zero of J_m, and $m = 0, 1, 2, \ldots, n = 1, 2, \ldots$. The dispersion relation for the mnth mode is

$$\frac{\omega^2}{c^2} = \frac{\zeta^2{}_{mn}}{p^2} + \beta^2 \quad .$$

In the TE modes the longitudinal magnetic field G_3 has the same form (9.39), but the boundary condition now requires ζ_{mn} to be replaced by a zero of $J'_m(\zeta)$.

9.4.3 Co-axial cylindrical wave guide

When the cross-section of the wave guide is the annular region between the circles $r = p$ and $r = q$, the origin is excluded and we must therefore expect a contribution from the Bessel functions of the second kind Y_m. For TM modes the boundary conditions are of the form

$$C\,J_m(\gamma p) + D\,Y_m(\gamma p) = 0$$

$$C\,J_m(\gamma q) + D\,Y_m(\gamma q) = 0 \quad .$$

The allowed values of γ are therefore the solutions of the equation

$$J_m(\gamma p)\,Y_m(\gamma q) = J_m(\gamma q)\,Y_m(\gamma p) \quad .$$

In this case there are also TEM modes with $\gamma = 0$. (See Problem 9.8).

PROBLEMS

9.1 From Maxwell's equations (9.2) to (9.5) in the case of a uniform medium
(i) verify the equation of charge conservation $\dot{\rho} + \operatorname{div} \mathbf{j} = 0$,
(ii) Show that, if $\mathbf{E}$ satisfies (9.18) and (9.2) then $\mathbf{H}$ can be defined so that (9.3) to (9.5) are satisfied,
(iii) find the inhomogeneous wave equation for $\mathbf{H}$.

9.2 Show by taking the divergence of (9.5) that in a medium of conductivity σ the charge density ρ satisfies the equation $\sigma\rho + \epsilon\dot{\rho} = 0$, and deduce that any initial charge density decays exponentially.

9.3 Plane polarised harmonic waves of the form $I\,e^{i(kx-\omega t)}$ incident in region 1 fall normally on to the plane boundary $x = 0$ between region 1 and region 2. Region 2 is given by $0 < x < h$, and region 3 is $x > h$. The refractive indices of 1, 2 and 3 are 1, n, np respectively.

Write $$E = Ae^{i(nkx-\omega t)} + Be^{i(-nkx-\omega t)} \quad (0 < x < h)$$

and $$E = Te^{inpk(x-h)-i\omega t} \quad (x > h) \quad ,$$

explaining why these forms are chosen, and hence find the transmission coefficient from region 1 to region 3 for waves of length $2\pi/k$. By considering the amplitude of the wave reflected into region 1 show that 100% transmission is possible only if either (i) $np = 1$ or (ii) $n = p$. In both cases (i) and (ii) find the relation between h and the wavelength in region 2 for perfect transmission. [Case (ii) shows that the refractive index of an ideal lens coating should be the square root of that of the glass].

9.4 In the case where the incident electric field is perpendicular to the plane of incidence (section 9.3.3) show, by considering the Poynting vector in the region $x < 0$, that the rate at which energy crosses any plane surface $x = $ constant is the difference between the fluxes due to the incident and the reflected waves.

9.5 From the Fresnel formulae (9.31) show that if $\tan\theta = n$ (**Brewster's angle**) the reflection is polarised perpendicular to the plane of incidence.

9.6 Show that a plane polarised harmonic wave incident at an angle θ to a plane boundary and undergoing *total* reflection will in general become elliptically polarised. Show that this polarisation can be circular only if the incident wave is polarised at an angle $\pi/4$ to the plane of incidence and $n \leqslant \sqrt{2} - 1$.

9.7 Let D be the region bounded by Γ in the wave guide of section 9.4. Show that, for a TM mode, the average energy flow along the guide, which is defined as

$$\int_D \mathbf{N} \,.\, \hat{\mathbf{z}} \; dS, \text{ where } \mathbf{N} = \tfrac{1}{2}\mathbf{E} \times \bar{\mathbf{H}} \quad .$$

is

$$\frac{1}{2} \frac{\epsilon\beta\omega}{k^2 - \beta^2} \int_D |F_3|^2 \, dS \quad .$$

Find the energy residing in unit length of the guide, defined as $\int_D U dS$, where $U = \frac{1}{4}(\epsilon \; \mathbf{E} \,.\, \bar{\mathbf{E}} + \mu \mathbf{H} \,.\, \bar{\mathbf{H}})$, and deduce that energy is transported at the group velocity.

9.8 Show that the electric field in the TEM mode of the coaxial guide of section 9.4.3 has the form $\mathbf{E} = \hat{\mathbf{r}} \, A r^{-1} e^{ik(z-ct)}$. Find the magnetic field and use (9.13) to find the currents flowing in the conductors.

9.9 Let $\psi(\mathbf{r})$ be a function satisfying $\nabla^2\psi + k^2\psi = 0$.
Show that

$$\mathbf{E} = ik \text{ curl } \psi \mathbf{r} e^{-ikct}$$

$$\mathbf{H} = \epsilon c \text{ curl curl } \psi \mathbf{r} e^{-ikct}$$

satisfy Maxwell's equations.
Hence show that there are normal oscillations of the field inside a spherical cavity of radius a in a perfect conductor given by

$$\psi = \frac{\partial}{\partial z} \frac{\sin kr}{r} \quad ,$$

with frequencies given by the roots of the equation

$$ka = \tan ka \quad .$$

CHAPTER 10

Water Waves

10.1 EQUATIONS OF LIQUID MOTION UNDER GRAVITY

The different mechanical properties of gases and liquids result principally from their varying compressibility, in that gases are compressed much more easily than liquids. Consequently when considering the propagation of waves in a liquid we assume the liquid to be an incompressible fluid.

So, for liquids, we take a constant value of ρ in (8.2) and (8.4) and, as before, we linearise (8.4) by assuming that

$$|\mathbf{q}\,.\,\nabla\,\mathbf{q}| \ll |\mathbf{q}_t| \quad . \tag{10.1}$$

Equations (8.5) and (8.6) now become

$$\operatorname{div}\mathbf{q} = 0,\quad \frac{\partial \mathbf{q}}{\partial t} + \nabla\left(\frac{p}{\rho} + W\right) = 0 \quad .$$

A slight modification of the argument of section 8.2 then leads to (8.10):

$$\mathbf{v} = \nabla\,\phi, \quad \text{and hence to}$$

$$\nabla^2\phi = 0,\quad \frac{\partial \phi}{\partial t} + \frac{p}{\rho} + W = 0 \quad . \tag{10.2}$$

For continuous, non-turbulent, motion of a liquid it is assumed that a boundary surface always consists of the same particles of liquid. We deduce that, if $F(\mathbf{r},\ t) = 0$ be such a surface, then, following the motion of the liquid particles,

$$\frac{\partial F}{\partial t} + \mathbf{v}\,.\,\nabla F = 0 \quad . \tag{10.3}$$

Finally we note that at a free surface of the liquid the surface tension effects produce a discontinuity in the pressure across the surface. That is,

$$p_1 = p_0 - T\left(\frac{1}{R_1} + \frac{1}{R_2}\right) \quad , \tag{10.4}$$

where p_1, p_0 are values of the pressure just inside and just outside the surface, T is the constant surface tension and R_1, R_2 are the principal radii of curvature of the surface.

To investigate the oscillations of a liquid we select Cartesian axes with the origin in the plane equilibrium surface of the liquid, the y-axis vertical and the xz-plane horizontal. The elevation of the liquid above the point $(x, 0, z)$ is denoted by $\eta(x, z, t)$, so the equation of the free surface is

$$\eta(x, z, t) - y = 0 \quad .$$

Similarly the equation of the lower boundary over which the liquid moves is given by

$$h(x, z, t) + y = 0 \quad .$$

That is $h(x, z, t)$ is the depth of the liquid in the equilibrium state. (See Fig. 10.1.)

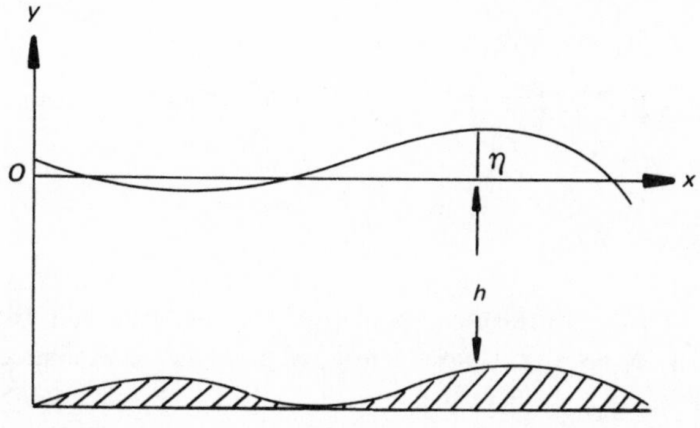

Fig. 10.1

Hence (10.3) gives

$$\nabla\phi \,.\, \nabla\eta - \phi_y = -\eta_t \text{ on } y = \eta \tag{10.5}$$

and

$$\nabla\phi \,.\, \nabla h + \phi_y = -h_t \text{ on } y = -h \quad . \tag{10.6}$$

For motion under gravity $W = gy$ and (10.2) gives

$$p = -\rho g y - \rho \frac{\partial\phi}{\partial t} \quad . \tag{10.7}$$

10.2 LINEARISATION OF THE BOUNDARY CONDITIONS

We now consider small motions of the boundaries, defined by the following approximations.

Approximation 1 The boundaries suffer only small departures from the horizontal, that is $|\nabla\eta| << 1$ and $|\nabla h| << 1$. This means that we may neglect the terms $\nabla\phi.\nabla\eta$ and $\nabla\phi.\nabla h$ in (10.5) and (10.6). Also in (10.4) the factor $\dfrac{1}{R_1}+\dfrac{1}{R_2}$ may be replaced by $\dfrac{\partial^2\eta}{\partial x^2}+\dfrac{\partial^2\eta}{\partial z^2}$.

Approximation 2 The elevation η is small enough to neglect changes in the vertical component of the velocity **v** between the free surface and the equilibrium plane. That is $\eta << \left(\dfrac{\phi_y}{\phi_{yy}}\right)_{y=0}$,

and as $$\phi_y(x, \eta, z, t) \cong \phi_y(x, 0, z, t) + \eta\phi_{yy}(x, 0, z, t)$$

the kinematical condition (10.5) reduces to

$$\phi_y = \eta_t \text{ on } y = 0 \quad . \tag{10.8}$$

Similarly if $|h - h_0| << \left(\dfrac{\phi_y}{\phi_{yy}}\right)_{y=-h_0}$, where h_0 is the mean depth of the fluid, (10.6) reduces to

$$\phi_y = -h_t \text{ on } y = -h_0 \quad . \tag{10.9}$$

Approximation 3 The vertical component of the acceleration is small compared with g, that is, $\phi_{ty} << g$. Consequently, when (10.7) is evaluated on the free surface we have

$$\begin{aligned} p(x, \eta, z, t) &= -\rho g\eta - \rho\phi_t(x, \eta, z, t) \\ &\cong -\rho g\eta - \rho\phi_t(x, 0, z, t) - \rho\eta\phi_{ty}(x, 0, z, t) \\ &\cong -\rho g\eta - \rho\phi_t(x, 0, z, t) \quad . \end{aligned}$$

Combining this with (10.4), and absorbing the constant atmospheric pressure p_0 into the velocity potential ϕ, gives

$$\phi_t + g\eta - \frac{T}{\rho}\nabla^2\eta = 0 \text{ on } y = 0 \quad . \tag{10.10}$$

10.3 ONE-DIMENSIONAL WAVES ON A HORIZONTAL BED

10.3.1 Solution of the linearised equations

When all the variables are independent of the z-coordinate and h is constant, the problem takes the form:
Find functions $\phi(x, y, t)$ and $\eta(x, t)$ satisfying

$$\phi_{xx} + \phi_{yy} = 0 \quad , \tag{10.11}$$

$$\phi_y(x, 0, t) = \eta_t(x, t) \quad , \tag{10.12}$$

$$\phi_y(x, -h, t) = 0 \quad , \tag{10.13}$$

and

$$\rho\phi_t(x, 0, t) + \rho g\eta - T\eta_{xx} = 0 \quad . \tag{10.14}$$

As (10.11) is separable we look for progressive wave solutions of the form

$$\phi = Y(y) \cos(kx - \omega t) \quad .$$

Then

$$Y''(y) = k^2 Y(y)$$

and so

$$\phi = A \cosh k(y + \epsilon) \cos (kx - \omega t) \quad ,$$

where A and ϵ are constants.
The boundary condition (10.13) gives $\epsilon = h$, and by using (10.12) we obtain $\eta_t = Ak \sinh hk \cos(kx - \omega t)$.
Hence

$$\eta = a \sin(kx - \omega t) \quad , \tag{10.15}$$

where a, the surface wave amplitude, is constant, and

$$\phi = -\frac{\omega a}{k \sinh kh} \cosh k(y + h) \cos(kx - \omega t) \quad . \tag{10.16}$$

Inserting (10.15) and (10.16) into (10.14) gives the dispersion relation

$$\omega^2 = gk\left(1 + \frac{T}{\rho g} k^2\right) \tanh kh \quad . \tag{10.17}$$

The observed pattern of the surface waves resulting from this relation has already been discussed in section 5.5.

Similar solutions may be derived for waves travelling in the opposite direction or for standing waves of the form $\phi = Y(y) \cos kx \cos \omega t$.

At this stage we should determine whether the solutions (10.15) and (10.16) are small motions, that is, conforming to the condition (10.1) and to the three approximations of section 10.2. It can be shown (see Problem 10.1) that all the conditions are satisfied if both $ak \ll 1$ and $a \ll h$, that is, if the amplitude of the surface oscillations is small compared with both the wavelength and the mean depth of the liquid.

The incompressibility assumption may be tested by solving the set of linearised equations with (10.11) replaced by

$$\phi_{xx} + \phi_{yy} = \frac{1}{c^2}\phi_{tt} \quad ,$$

where c is the speed of sound in the liquid. A similar analysis gives the solution (10.15) together with

$$\phi = -\frac{\omega a}{p \sinh ph} \cosh p(y+h) \cos(kx - \omega t)$$

where

$$p^2 = k^2(1 - \frac{V^2}{c^2}) \text{ and } V = \frac{\omega}{k}$$

Thus the assumption is justified for particle speeds $V \ll c$. For water at 15°C, c = 1470 m/sec and so normal oscillations in water are most unlikely ever to be influenced by compressibility.

10.3.2 Motion of individual particles

To find the individual particle motion we consider a particle whose position in the equilibrium state is (x_0, y_0) and let its displaced position at any subsequent time be $[x_0 + \alpha(t), y_0 + \beta(t)]$. Then using (10.16),

$$\frac{d\alpha}{dt} \approx \phi_x(x_0, y_0, t) = \frac{\omega a}{\sinh kh} \cosh k(y_0 + h) \sin(kx_0 - \omega t)$$

and

$$\frac{d\beta}{dt} \approx \phi_y(x_0, y_0, t) = -\frac{\omega a}{\sinh kh} \sinh k(y_0 + h) \cos(kx_0 - \omega t) \quad .$$

Integrating with respect to t gives,

$$\alpha = \frac{a}{\sinh kh} \cosh k(y_0 + h) \cos(kx_0 - \omega t)$$

and

$$\beta = \frac{a}{\sinh kh} \sinh k(y_0 + h) \sin(kx_0 - \omega t) \quad .$$

Thus the disturbed motion of an individual particle is an ellipse with horizontal and vertical axes. At the bottom of the liquid, when $y_0 = -h$, the ellipse degenerates into a straight line and the particles oscillate horizontally.

When $kh \gg 1$, the wavelength is small compared with the depth, and the particle motion is almost circular. The radius of the circle decreases very rapidly below the surface (consider a depth of one wavelength, $y_0 = -2\pi/k$), and so such waves are often called **surface waves.** When $kh \ll 1$, the wavelength is large compared with the depth, and the motion is almost entirely horizontal and independent of y_0. These are plane longitudinal waves, and are often called **tidal waves** because tidal movements, which have very low frequency, are examples of this type of motion.

The particle motion for standing waves is discussed in Problem 10.2 where it is shown that the motion of each particle is rectilinear and simple harmonic; the direction of motion being vertical under the crests and troughs and horizontal under the nodes.

10.3.3 Solutions not influenced by surface tension

We see from (10.17) that the surface tension has a negligible influence for wave numbers $k \ll k_m$, where $k_m^2 = \rho g T^{-1}$. Water with $T = 0.074$ Nm^{-1} and $\rho = 1000$ kg m^{-3} has $k_m = 360$m^{-1} and the corresponding wavelength $\lambda_m = 0.017$m. Waves whose wavelength $\lambda \gg \lambda_m$ are called pure gravity waves and in practice this term is applied to all water waves with $\lambda > 0.1$m.

For pure gravity waves the dispersion relation (10.17) reduces to

$$\omega^2 = gk \tanh kh \quad . \tag{10.18}$$

It follows that for **long waves**, or shallow water, when $kh \ll 1$,

$$\omega^2 = ghk^2$$

and all waves travel with the same speed $(gh)^{1/2}$. Such waves are not dispersed and so their displacements must satisfy the classical one-dimensional wave equation

$$\frac{\partial^2 \eta}{\partial x^2} = \frac{1}{V^2}\frac{\partial^2 \eta}{\partial t^2}, \quad \text{where } V^2 = gh \quad .$$

Consequently these wave motions may be analysed by the methods described in previous chapters.

For **short waves**, or deep water, when $kh \gg 1$, the dispersion relation (10.18) reduces to

$$\omega^2 = gk \quad .$$

Thus the phase speed $V = \sqrt{(g/k)}$ and the group speed $U = \frac{1}{2}V$. So V is proportional to the square root of the wavelength and is independent of the depth.

The limit $kh \gg 1$ may be clarified somewhat, for it implies the approximation $\tanh kh \approx 1$. Standard tables show that $\tanh 5 = 0.99991$ and so to this degree of accuracy we require $kh > 5$, or $h > 0.8\lambda$. This limiting case well represents ocean waves created by wind.

Finally we note that for gravity waves the pressure at any point in the liquid can be calculated from (10.7), the velocity potential and the now assumed continuity of pressure across the free surface. For long waves on shallow water ($kh \ll 1$) it is easily seen that

$$p = p_0 + g\rho(\eta - y)$$

so that the pressure at any point is equal to the hydrostatic pressure head at that point; a fact which follows immediately from the result of section 10.3.2 that the vertical motion is negligible in this case. Similarly for waves on deep water ($kh \gg 1$) it is found that the pressure at any fluid particle position is equal to that at the undisturbed particle position.

10.4 PARTICULAR ONE-DIMENSIONAL FLOWS

10.4.1 Surges in a uniform channel

In certain circumstances a natural surge, or **bore**, may be created by tidal action in large estuaries and rivers. A positive surge wave is one which results in an increase in the depth of the liquid whilst a negative surge produces a decrease in the depth.

We consider the propagation of a surge wave in a uniform channel of rectangular cross-section and horizontal base. It is assumed that the particle speed is constant over a given cross-section and that a uniform flow with speed q_1 and depth h_1 is disturbed by a positive surge moving upstream with constant speed c relative to the bottom of the channel. Then a short distance downstream of the surge the flow is again uniform with speed q_2 and depth h_2.

To investigate the motion we select coordinate axes moving with the wave, so that the wave appears stationary and the uniform flows have speeds $v_1 = q_1 + c$ and $v_2 = q_2 + c$, as in Fig. 10.2. As this is now a steady

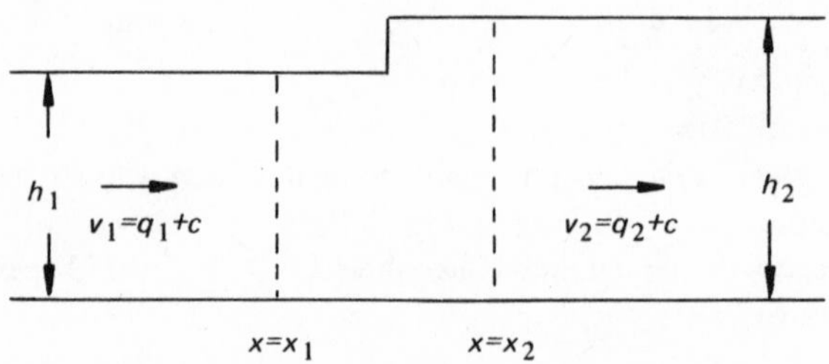

Fig. 10.2

flow system we may apply the momentum equation to the mass of liquid which, at a given instant, is contained between two cross-sections $x = x_1$ and $x = x_2$, one on each side of the discontinuity. As the thrust on each section is hydrostatic the net horizontal force on the mass is $\frac{1}{2}\rho gb(h_1^2 - h_2^2)$, where b is the width of the channel. The rate of increase of momentum of the mass is $\rho Q(v_2 - v_1)$, where Q is the rate at which volume of liquid crosses each section of the channel.

Thus $$Q = v_1 h_1 b = v_2 h_2 b$$

and $$\tfrac{1}{2}\rho gb(h_1^2 - h_2^2) = \rho v_1 h_1 b(v_2 - v_1) \quad .$$

Hence $$v_1 = \left\{\frac{gh_2(h_1 + h_2)}{2h_1}\right\}^{1/2}$$

and this is the speed of the surge wave relative to the liquid upstream of the discontinuity.

Now consider the energy of the liquid bounded by the planes $x = x_1$ and $x = x_2$. The rate of decrease of kinetic energy is

$$\tfrac{1}{2}\rho b v_1 h_1(v_1^2 - v_2^2) \quad ,$$

the rate of decrease of potential energy is

$$-\tfrac{1}{2}\rho bg v_1 h_1(h_2 - h_1)$$

and the rate of working of the pressure forces over the sections is

$$\int_0^{h_1} (p_0 + \rho gy) b v_1\, dy - \int_0^{h_2} (p_0 + \rho gy) b v_2\, dy = -\tfrac{1}{2}\rho bg v_1 h_1(h_2 - h_1) \quad .$$

So the total rate of release of energy is

$$\frac{\rho bg\, v_1(h_2 - h_1)^3}{4h_2} \quad .$$

Hence energy is not conserved, unless $h_1 = h_2$ when the surge does not exist. It follows that, in a negative bore ($h_2 < h_1$), an external supply of energy is required if the stable system is to be maintained. Consequently a negative bore of finite amplitude cannot be propagated with a constant profile. Likewise a positive bore ($h_2 > h_1$) can only be maintained if there is some mechanism whereby the excess energy can be dissipated. In practice the energy losses are often accounted for by a turbulent region in the transition from depth h_1 to depth h_2.

10.4.2 Channel flow over a small obstacle

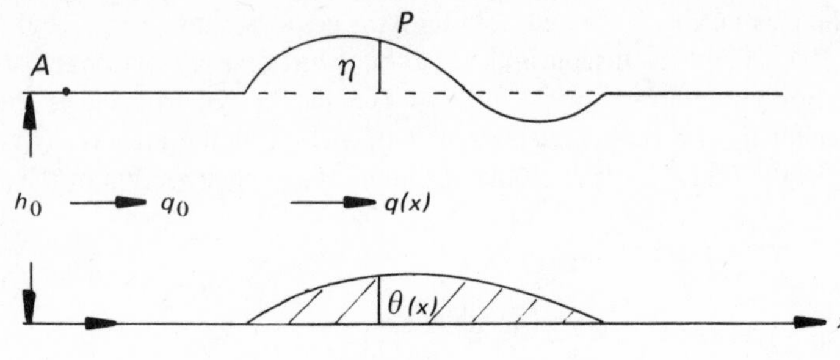

Fig. 10.3

We now consider how static waves may be formed on the free surface of a liquid in a channel of constant width when uniform flow of speed q_0 and depth h_0 encounters a small hump in the horizontal bed (see Fig. 10.3). Let $\theta(x)$ be the height of the hump above the channel bed, and suppose that a steady, approximately horizontal flow $q(x)\mathbf{i}$ exists in which $\eta(x)$ is the elevation of the surface. We assume that θ and η are small enough to satisfy the linearisation assumptions of section 10.2, but q is not small. Then, from continuity,

$$q_0 h_0 = q(h_0 + \eta - \theta) \quad . \tag{10.19}$$

The Euler equation (8.4) simplifies to

$$q\frac{\mathrm{d}q}{\mathrm{d}x}\,\mathbf{i} + \frac{1}{\rho}\,\nabla p + \nabla W = 0 \quad ,$$

which can be integrated to give

$$\tfrac{1}{2}q^2 + \frac{p}{\rho} + W = \text{constant} \quad . \tag{10.20}$$

Evaluating this at A and P in Fig. 10.3, noting that at both points p is the atmospheric pressure, and taking $W = 0$ at A, we have

$$\tfrac{1}{2}q_0^2 = \tfrac{1}{2}q^2 + g\eta \quad . \tag{10.21}$$

Eliminating q from (10.19) and (10.21) gives

$$\frac{2g\eta}{q_0^2} = 1 - \left(1 + \frac{\eta - \theta}{h_0}\right)^{-2} \quad . \tag{10.22}$$

Assuming that $\theta << h_0$ and $\eta << h_0$ we may write

$$\left(1+\frac{\eta-\theta}{h_0}\right)^{-2} \cong 1-\frac{2(\eta-\theta)}{h_0}$$

and conclude that
$$\eta(x)=\frac{\theta(x)}{1-F^{-1}}, \tag{10.23}$$

where $F=\dfrac{q_0^2}{gh_0}$ is the **Froude number.**

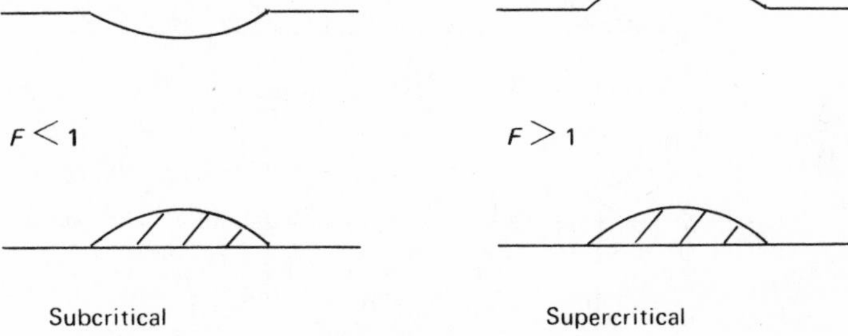

Fig. 10.4

Hence the surface profile is geometrically similar to the bottom profile, the size of the image depending on the Froude number. When $q_0^2 < gh_0$, $F < 1$, and the flow is said to be **subcritical.** The image is inverted, and corresponding to the hump there is a static wave in the form of a depression in the free surface. When $q_0^2 > gh_0$, $F > 1$ and the flow is **supercritical.** The image is upright, and the hump produces an elevation in the free surface. The two types of flow are illustrated in Fig. 10.4. For the critical case $F = 1$, when the flow speed q_0 is equal to the speed of long waves in the channel, equation (10.23) provides no solution and our simple analysis ceases to be valid. An investigation of this case is beyond the scope of this book.

10.5 GRAVITY WAVES ON THE INTERFACE OF TWO LIQUIDS

We now consider the propagation of gravity waves along the common surface of two superposed liquids. With the coordinate axes of section 10.1 we suppose that in the undisturbed flow liquids of densities ρ_1, ρ_2 occupy the regions $0 < y < h_1$ and $-h_2 < y < 0$ respectively, where $y = h_1$ and $y = -h_2$ are rigid boundaries. Suppose that the liquids are flowing uniformly with speeds u_1 and u_2 in the direction of the x-axis.

Now let us consider a small perturbation of this steady flow. Let the velocity potential be $\phi = \phi_1$ $(y > 0)$ and $\phi = \phi_2$ $(y < 0)$, and let the interface have the form $y = \eta(x, t)$ (see Fig. 10.5). The unperturbed values are $\phi_1 = u_1 x$, $\phi_2 = u_2 x$, $\eta = 0$. We shall assume that the perturbations of ϕ and η are small enough to satisfy the linearisation assumptions of section 10.2. Then, as in section 10.3.1, we require

$$\phi_{xx} + \phi_{yy} = 0, \quad \phi_y(x, h_1, t) = 0, \quad \phi_y(x, -h_2, t) = 0 \quad .$$

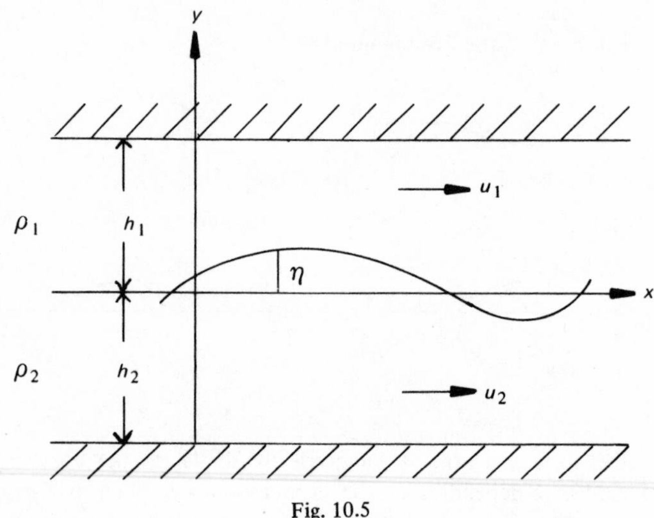

Fig. 10.5

The kinematical boundary condition (10.3) applied at the interface $\eta - y = 0$ gives

$$\frac{\partial \eta}{\partial t} + u_1 \frac{\partial \eta}{\partial x} = \frac{\partial \phi_1}{\partial y} \text{ and } \frac{\partial \eta}{\partial t} + u_2 \frac{\partial \eta}{\partial x} = \frac{\partial \phi_2}{\partial y} \quad \text{on } y = 0 \quad .$$

When the velocity is approximately u in the x-direction the Euler equation (8.4) takes the form

$$q_t + u q_x + \rho^{-1} \nabla p + \nabla W = 0 \quad ,$$

giving $$\phi_t + u\phi_x + \rho^{-1} p + W = \text{ constant} \quad .$$

If the surface tension is negligible the pressure must be continuous across the interface, and so

$$\rho_1 \left(\frac{\partial \phi_1}{\partial t} + u_1 \frac{\partial \phi_1}{\partial x} + g\eta \right) = \rho_2 \left(\frac{\partial \phi_2}{\partial t} + u_2 \frac{\partial \phi_2}{\partial x} + g\eta \right) \quad \text{on } y = 0 \quad .$$

Extending the treatment of section 10.3.1, we seek harmonic wave solutions

$$\phi_1 = u_1 x + A_1 \cosh k(y - h_1)\, e^{i(kx - \omega t)} \quad ,$$

$$\phi_2 = u_2 x + A_2 \cosh k(y + h_2)\, e^{i(kx - \omega t)} \quad ,$$

$$\eta = iae^{i(kx - \omega t)} \quad .$$

Substituting these into the equations and eliminating the constants A_1, A_2 and a yields the dispersion relation

$$kg(\rho_2 - \rho_1) = \rho_1(u_1 k - \omega)^2 \coth kh_1 + \rho_2(u_2 k - \omega)^2 \coth kh_2 \quad . \tag{10.24}$$

In the simplified case of deep liquids (or short waves), $kh_1 \gg 1$ and $kh_2 \gg 1$, and (10.24) reduces to

$$kg(\rho_2 - \rho_1) = \rho_1(u_1 k - \omega)^2 + \rho_2(u_2 k - \omega)^2 \quad . \tag{10.25}$$

It is now convenient to introduce the reference frame in which the total momentum is zero, which moves with velocity $(\rho_1 + \rho_2)^{-1}(\rho_1 u_1 + \rho_2 u_2)$. With respect to this frame the stream velocities are

$$U_1 = -\frac{\rho_2 U}{\rho_1 + \rho_2} \,, \qquad U_2 = \frac{\rho_1 U}{\rho_1 + \rho_2} \,, \qquad \text{where } U = u_2 - u_1 \quad .$$

Then (10.25) reduces to

$$(\rho_1 + \rho_2)^2 \omega^2 + \rho_1 \rho_2 U^2 k^2 = (\rho_2^2 - \rho_1^2) gk \quad .$$

The resulting dispersion curve is an arc of an ellipse (Fig. 10.6).

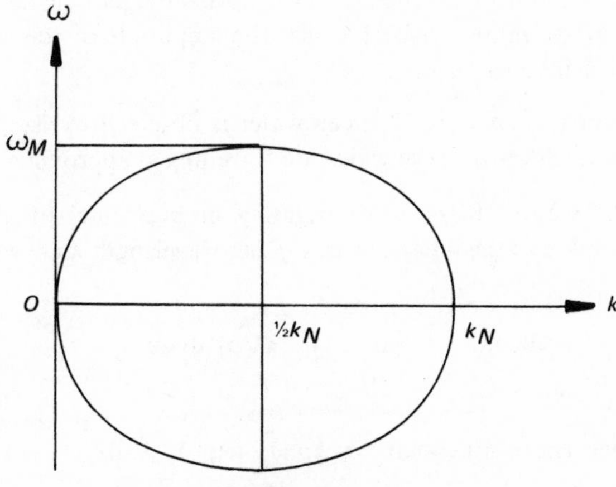

Fig. 10.6

It is immediately clear that the system will only transmit undistorted, unattenuated harmonic waves when ω and k satisfy

$$|\omega| < \omega_M \text{ and } 0 \leqslant k \leqslant k_N \quad ,$$

where

$$\omega_M = \frac{\rho_2 - \rho_1}{2\sqrt{(\rho_1\rho_2)}} \frac{g}{U} \text{ and } k_N = \frac{\rho_2^2 - \rho_1^2}{\rho_1\rho_2} \frac{g}{U^2} \quad .$$

For $k > k_N$, ω is pure imaginary and the time dependence of η is exponential, so that there can be no oscillations. When $\omega > \omega_M$, k has the form $k = \frac{1}{2}k_N \pm i\mu$, so that the waves may be attenuated or may increase and become unstable. The latter case must be treated as a non-linear problem and is beyond the scope of this book.

PROBLEMS

10.1 Prove that the solutions (10.15) and (10.16) are consistent with the approximations $|\mathbf{q} \cdot \nabla\mathbf{q}| << |\mathbf{q}_t|$, $|\nabla\eta| << 1$, $\eta << (\phi_y/\phi_{yy})_{y=0}$ and $(\phi_{ty})_{y=0} << g$ provided both $ak << 1$ and $a << h$.

10.2 Prove that there exist standing wave solutions of the linearised equations (10.11) – (10.14) of the form

$$\eta = a\sin kx \cos\omega t,\ \phi = -\frac{ga}{\omega} \frac{\cosh k(h+y)}{\cosh kh} \sin kx \sin\omega t \quad .$$

Deduce that the disturbed motion of each particle is rectilinear and simple harmonic, the direction of motion being vertical under the crests and troughs and varying to horizontal under the nodes. Show also that as we pass from the free surface to the bottom of the liquid the horizontal particle motion decreases its amplitude in the ratio $\cosh kh{:}1$ whilst the amplitude of the vertical motion decreases from $a\sin kx$ to zero.

10.3 A boat which is anchored in deep water is observed to rise and fall fifteen times in a minute. Show that the waves are travelling at approximately 23 km/hr.

10.4 When the wave (10.16) is propagating in a channel of unit width the potential energy V and the kinetic energy T per wavelength λ are given by

$$V = \tfrac{1}{2}\rho g \int_0^\lambda \eta^2 \mathrm{d}x \text{ and } T = \tfrac{1}{2}\rho \int_0^\lambda \int_{-h}^{\eta} (\nabla\phi)^2 \mathrm{d}y\,\mathrm{d}x \quad .$$

Use Green's theorem to show that $T = \frac{1}{2}\rho \displaystyle\int_0^\lambda (\phi\phi_y)_{y=0}\, \mathrm{d}x$.

Show that $T = V = \frac{1}{4}\rho g\, a^2\lambda$.

10.5 For the system in Problem 10.4 the rate of transmission of energy across any cross-section of the channel is given by W, where

$$W = \int_{-h}^{0} p\phi_x \mathrm{d}y \quad .$$

Use equation (10.7) to substitute for p and show that, neglecting surface tension, the *mean value* of W over a complete cycle is

$$\bar{W} = \tfrac{1}{4}\rho g a^2 \, \frac{\omega}{k} \left(1 + 2kh \text{ cosech } 2kh\right) \quad .$$

Deduce that the speed of the energy flow is equal to the group speed.
[Hint: The mean value of $\sin(kx - \omega t)$ is zero, and of $\sin^2(kx - \omega t)$ it is $\frac{1}{2}$].

10.6 In the undisturbed state water of uniform depth h is flowing with speed proportional to the distance from the bottom, u being the speed of the stream at the surface. Prove that when gravity waves are propagated in the direction of the stream the phase speed V is given by

$$gh(V-u)^2 + u(V-u)\,V_0^2 = V_0^2 gh \quad ,$$

where V_0 is the phase speed of waves in still water.

10.7 A deep horizontal canal of rectangular cross-section is closed at its ends by two fixed vertical barriers at a distance L apart. Find the periods of oscillation of the normal modes and show that in the nth mode the particles at depth d below the surface all oscillate in straight lines of length proportional to $e^{-n\pi d/L}$.

10.8 A canal of length L and uniform depth h has a width which varies uniformly with distance from the end $x = 0$. Find the surface profile at any instant given that the end $x = L$ meets an open sea in which a tidal oscillation $\eta(L, t) = b\cos\omega t$ is maintained.

10.9 A stream of liquid of mean depth h is flowing with uniform speed u over a corrugated bed. When the bed has a profile given by

$$y = -h + b \sin kx, \text{ where } b \ll h \quad ,$$

show that the static wave on the free surface is

$$\eta = \frac{a \sin kx}{\cosh kh - (g/ku^2)\sinh kh} \quad .$$

10.10 The bottom of a horizontal canal of mean depth h is agitated so that its form is given by

$$y = -h + b\cos(kx - \omega t), \text{ where } b \ll h \quad .$$

Show that the form of the free surface is given by

$$\eta = \frac{b\omega^2\cos(kx - \omega t)}{\omega^2\cosh kh - kg\sinh kh}$$

10.11 A horizontal pipe is completely filled by two liquids. In the equilibrium configuration liquid of density ρ_1 occupies a cross-sectional area A_1 and liquid of density ρ_2 an area A_2. The liquids have a common horizontal surface of width b and when long waves travel on the surface the speed of propagation is V. Impose a speed $-V$ on the system in order to produce static waves and then apply the equation of continuity to each liquid separately. Hence, after using equation (10.20) and the continuity of pressure across the interface, show that

$$V^2 = \frac{g(\rho_2 - \rho_1)A_1A_2}{b(A_1\rho_2 + A_2\rho_1)} \quad .$$

10.12 Two liquids of densities ρ_1, ρ_2 have a horizontal surface of separation but are otherwise unbounded. Show that when small amplitude waves are propagated on the common surface the individual particles of the two fluids describe circles about their mean positions.

10.13 Air of density ρ_1 blows over a deep stationary liquid of density ρ_2 and the speed of the air is just sufficient to prevent waves propagating against it. Show that the speed of propagation of waves with the air is $2c\,\{\rho_1/(\rho_1 + \rho_2)\}^{1/2}$, where c is the wave speed when the air is stationary.

10.14 A liquid of density ρ_1 and depth h_1 rests on a second liquid of density ρ_2 and depth h_2, so that in equilibrium the common surface and the free surface of the upper liquid are both horizontal. Show that when waves of the same frequency are propagated on the free surface and along the interface the ratio of their respective amplitudes is $1:\{\cosh kh_1 - (gk/\omega^2)\sinh kh_1\}$, and the dispersion relation is

$$\omega^4(\rho_1 + \rho_2\coth kh_1\coth kh_2) - \omega^2 gk\rho_2(\coth kh_1 + \coth kh_2) + g^2k^2(\rho_2 - \rho_1) = 0.$$

10.15 If in Problem 10.11 the liquids are moving with mean horizontal speeds

u_1 and u_2, and there is a surface tension T, prove that the wavelength λ is determined by

$$\frac{4T\pi^2}{\lambda^2} = b(\rho_1 u_1^2/A_1 + \rho_2 u_2^2/A_2) - g(\rho_2 - \rho_1) \quad .$$

10.16 Show that for surface waves on a deep liquid the phase velocity V and the group velocity U satisfy the equation

$$UV \approx \frac{g}{2k} + \frac{3Tk}{\rho} \quad .$$

10.17 Given that $T = 0.073$ Nm^{-1} for air over water and that $\rho = 1.22$ kg m^{-3} for air, find the speed of ripples of length 1cm on water surmounted by air.

10.18 The undisturbed surface of a deep liquid is the plane $y = 0$. A floating body is placed on the surface and it creates a shallow depression, bounded by the planes $y = \epsilon(\pm\, x/a - 1)$, where ϵ is small. Show that when the floating body is suddenly removed the equation of the free surface at a subsequent time t is given by

$$\eta = -\frac{4\epsilon}{a\pi} \int_0^\infty \frac{\cos kx \sin^2(\tfrac{1}{2}ka)\cos\sqrt{(gk)}t}{k^2}\, dk \quad .$$

CHAPTER 11

Theory of Hyperbolic Equations

11.1 THE CAUCHY PROBLEM

We now examine the general theory of partial differential equations which may be associated with wave propagation. In this chapter we consider one-dimensional equations with one dependent variable u and two independent variables x and t, where x is a space variable and t is time. Equations with more than one space variable are considered in Chapter 12.

The Cauchy problem may be formulated as follows:

> Given a partial differential equation of order k and a curve Γ, defined in a region of the xt-plane, find a solution for which u and all the partial derivatives up to order $k-1$ take some prescribed values on Γ.

The prescribed values of u and the derivatives are called the Cauchy data on Γ.

11.2 GENERAL SOLUTION OF FIRST ORDER EQUATIONS

Consider the first order quasi-linear equation

$$P(x, t, u)\frac{\partial u}{\partial x} + Q(x, t, u)\frac{\partial u}{\partial t} = R(x, t, u) \quad , \tag{11.1}$$

where P, Q, R are C^1 functions and $P \neq 0$, $Q \neq 0$ throughout the region under consideration. Any solution $u = u(x, t)$ of (11.1) represents a surface in xtu-space and is called an **integral surface.** Let A be an arbitrary point on an integral surface. As the tangent plane at A contains the vector (P, Q, R), the field line through A of the vector field $[(P, Q, R)]$ must lie in the integral surface. Such a field line is called a **characteristic curve** of (11.1), and its projection on to the xt-plane is called a **characteristic.** Since P, Q, R are assumed single valued there will be only one characteristic curve through A, and so the integral surface is generated by the characteristic curves. Now the curves are everywhere tangential to the vector (P, Q, R) and therefore, if

$$x = x(s), \quad t = t(s), \quad u = u(s)$$

are the parametric equations of a characteristic curve, then

$$\frac{dx/ds}{P} = \frac{dt/ds}{Q} = \frac{du/ds}{R} \quad . \tag{11.2}$$

As (11.2) comprises two independent ordinary differential equations its general solution has the form

$$\phi(x, t, u) = c_1, \qquad \psi(x, t, u) = c_2 \quad , \tag{11.3}$$

where c_1, c_2 are constants. Since any one-parameter family of characteristic curves generates an integral surface, there is an integral surface corresponding to every relation of the form

$$F(c_1, c_2) = 0 \quad . \tag{11.4}$$

Hence the general solution of (11.1) is

$$F(\phi, \psi) = 0 \quad , \tag{11.5}$$

where F is an arbitrary function.

We have derived this result by geometrical reasoning. It can be verified analytically by substituting the solution into the equation (see Problem 11.1).

Example
Find the general solution of the equation

$$t^2 u_x - xtu_t = x(u - 2t) \quad .$$

Solution
First consider the subsidiary equations

$$\frac{dx/ds}{t} = \frac{dt/ds}{-xt} = \frac{du/ds}{x(u-2t)} \quad .$$

It is easily seen that

$$x\frac{dx}{ds} + t\frac{dt}{ds} = 0$$

and
$$(u-t)\frac{dt}{ds} + t\frac{d}{ds}(u-t) = 0 \quad .$$

Hence two independent solutions of the subsidiary equations are

$$x^2 + t^2 = c_1 \quad \text{and} \quad t(u - t) = c_2 \quad .$$

Thus the general solution of the given equation is

$$F(x^2 + t^2,\ tu - t^2) = 0 \quad ,$$

or $\quad u = t + t^{-1}f(x^2 + t^2) \quad ,$

where F and f are are arbitrary functions.

11.3 THE CAUCHY PROBLEM FOR FIRST ORDER EQUATIONS

We now consider the problem of finding a solution of (11.1) which takes prescribed values of u on a curve Γ in the xt-plane. This is equivalent to finding an integral surface of (11.1) which contains a given curve C in the xtu-space. It may be shown (see Courant & Hilbert 1962, Ch. II) that

(i) when Γ intersects every characteristic once and only once the problem has a unique solution,
(ii) when C coincides with a characteristic curve the problem has infinitely many solutions,
(iii) if Γ is a characteristic but C is not a characteristic curve there is no solution,
(iv) if Γ touches a characteristic at a point T, then Γ will intersect some characteristics twice, and therefore the Cauchy conditions may in general be specified only on an interval of Γ which does not contain T as an interior point.

Consider the equation

$$tu_x - xu_t = 0 \quad .$$

The subsidiary equations have integrals $x^2 + t^2 = c_1$ and $u = c_2$. Thus the characteristics are circles in the xt-plane with centre at the origin, and the characteristic curves are the projections of the characteristics on to planes u = constant. The general solution of the equation is $u = f(x^2 + t^2)$, where f is arbitrary, and so the integral surfaces are surfaces of revolution about the u-axis.

We now impose different sets of Cauchy data.

(i) $u = x^2 + 2x^4$ on $t = 0, x \geqslant 0$: The curve Γ has just one intersection with each characteristic. We have $x^2 + 2x^4 = f(x^2)$. Thus $f(\lambda) = \lambda + 2\lambda^2$ and there is a unique solution

$$u = (x^2 + t^2) + 2(x^2 + t^2)^2 \quad .$$

(ii) $u = 1$ on $x^2 + t^2 = 1$: The curve C coincides with a characteristic curve. We have $f(1) = 1$, and any function f satisfying this condition provides a solution to the problem. Hence there is an infinite number of solutions.

(iii) $u = x^3$ on $x^2 + t^2 = 1$: Γ is a characteristic but C is not a characteristic curve. The function f must satisfy $f(1) = x^3$ when $x^2 + t^2 = 1$, and so no solution exists.

(iv) $u = x^2 + 2x^4$ on $t = 1$: Γ touches the characteristic $x^2 + t^2 = 1$ at the point $T(t = 1, x = 0)$, and has two intersections with all characteristics $x^2 + t^2 = a^2 (a > 1)$. Hence the Cauchy data must be restricted to a suitable interval of Γ, e.g. $x \geqslant 0$. Then we have $x^2 + 2x^4 = f(x^2 + 1)$,

whence $$f(y) = (y-1)^2 + 2(y-1)^4$$

and hence $u = (x^2 + t^2 - 1)^2 + 2(x^2 + t^2 - 1)^4$.
This solution is unique over the part of the plane outside the circle $x^2 + t^2 = 1$. However, for $x^2 + t^2 < 1$ no characteristic intersects Γ, and so the solution is not completely defined in this region.

11.4 CANONICAL FORMS OF SECOND ORDER EQUATIONS

In analytical geometry equations of the second degree in x and y may be reduced to a normal, or canonical, form by introducing new variables in which the equation assumes the standard form of a hyperbola, or ellipse or parabola referred to principal axes.

We now show that a similar reduction may be effected for the second order linear partial differential equation

$$au_{xx} + 2bu_{xy} + cu_{yy} + du_x + eu_y + fu = g \quad , \tag{11.6}$$

where $a, b, \ldots, g$ are C^1 functions of x and y. (It is more convenient to denote the independent variables by x and y at this stage, and to revert to x and t when applying the theory to wave propagation.) This equation is said to be **hyperbolic** if $ac < b^2$, **elliptic** if $ac > b^2$, and **parabolic** if $ac = b^2$.

In terms of new variables $\xi = \xi(x, y)$, $\eta = \eta(x, y)$ (11.6) becomes

$$AU_{\xi\xi} + 2BU_{\xi\eta} + CU_{\eta\eta} + DU_\xi + EU_\eta + FU = G \quad ,$$

where $u(x, y) = U(\xi, \eta)$ and $A, B, \ldots, G$ are functions of ξ and η. In particular,

$$A = a\,\xi_x^2 + 2b\,\xi_x\xi_y + c\,\xi_y^2 \quad , \tag{11.7}$$

$$B = a\,\xi_x\eta_x + b(\xi_x\eta_y + \xi_y\eta_x) + c\,\xi_y\eta_y \quad , \tag{11.8}$$

and $$C = a\,\eta_x^2 + 2b\,\eta_x\eta_y + c\,\eta_y^2 \quad . \tag{11.9}$$

It is easily verified that

$$AC - B^2 = (ac - b^2)\,[J(\tfrac{\xi\eta}{xy})]^2 \quad , \tag{11.10}$$

where $J\,(\frac{\xi\eta}{xy}) = \begin{vmatrix} \xi_x \xi_y \\ \eta_x \eta_y \end{vmatrix}$ is a Jacobian, and so the hyperbolic, elliptic, or parabolic character of the equation is unaffected by transformation of the independent variables.

11.4.1 Hyperbolic equations

When $ac < b^2$ the roots λ_1, λ_2 of the equation

$$a\lambda^2 + 2b\lambda + c = 0 \tag{11.11}$$

are real and distinct.

Choosing ξ, η so that

$$\xi_x = \lambda_1 \xi_y \text{ and } \eta_x = \lambda_2 \eta_y \tag{11.12}$$

we see that $A = C = 0$ and $B \neq 0$. Thus (11.6) reduces to the canonical form

$$U_{\xi\eta} = \Phi(\xi, \eta, U, U_\xi, U_\eta) \quad .$$

The curves $\xi =$ constant and $\eta =$ constant are called the **characteristics** of (11.6), and as $\lambda_1 \neq \lambda_2$ the curves cannot be tangential at any point. Thus the characteristics form a curvilinear coordinate system in the xy-plane. The variables ξ, η are called characteristic variables or characteristic coordinates.

11.4.2 Elliptic equations

When $ac > b^2$ the roots λ_1, λ_2 are complex conjugates and the solutions of (11.12) are complex conjugate functions

$$\xi = \phi + i\psi, \eta = \phi - i\psi \quad ,$$

where $\phi(x, y)$ and $\psi(x, y)$ are real. Consequently an elliptic equation has no real characteristic curves. To find a real canonical form we use the variables ϕ, ψ in which case (11.6) reduces to the form

$$U_{\phi\phi} + U_{\psi\psi} = \Phi(\phi, \psi, U, U_\phi, U_\psi) \quad .$$

11.4.3 Parabolic equations

When $ac = b^2$ we have $\lambda_1 = \lambda_2$ in which case (11.12) gives a single transformation

$\xi = \xi(x, y)$ which simultaneously gives $A = B = 0$. Consequently if $\eta = \eta(x, y)$ is any function independent of ξ, we have the canonical form

$$U_{\eta\eta} = \Phi(\xi, \eta, U, U_\xi, U_\eta) \quad .$$

11.4.4 Example
Reduce the equation

$$yu_{xx} + (x + y)u_{xy} + xu_{yy} = x + y$$

to canonical form.

Solution
As $b^2 - ac = \frac{1}{4}(x - y)^2$ the equation is hyperbolic everywhere except on the line $x = y$, where it is parabolic. As the roots of $y\lambda^2 + (x + y)\lambda + x = 0$ are $\lambda_1 = -x/y$ and $\lambda_2 = -1$ the characteristic coordinates ξ, η satisfy

$$y\xi_x + x\xi_y = 0 \text{ and } \eta_x + \eta_y = 0 \quad .$$

Hence $\xi = x^2 - y^2$ and $\eta = x - y$, and the given equation reduces to

$$\frac{\partial^2 U}{\partial\xi\partial\eta} + \frac{1}{\eta}\frac{\partial U}{\partial\xi} + \frac{\xi}{2\eta^3} = 0 \quad .$$

11.5 THE CAUCHY PROBLEM FOR SECOND ORDER EQUATIONS

We now restrict our attention to the equations which are naturally associated with wave propagation. We thus ignore elliptic equations which represent problems in mathematical physics concerned with equilibrium configurations and steady-state conditions which occur after a long time has elapsed. Hyperbolic and parabolic equations occur in problems where the state is known at time $t = 0$, and it is required to determine the state for $t > 0$. However it is hyperbolic equations which most naturally characterise problems in which an essential feature is a propagation of change of state as t increases. (See Problem 11.5.) Consequently we only consider hyperbolic equations.

11.5.1 Well-posed problems
When u and all its partial derivatives are known on a curve Γ it is possible to construct a power series solution of (11.6) valid in some neighbourhood of Γ. Suppose Γ is the curve

$$x = X(\sigma), y = Y(\sigma)$$

and that the Cauchy data on Γ are

$$u = L(\sigma),\ u_x = M_1(\sigma),\ u_y = M_2(\sigma) \quad .$$

Of course the three conditions are not independent, for

$$M_1(\sigma)\,X'(\sigma) + M_2(\sigma)\,Y'(\sigma) = L'(\sigma) \quad ,$$

which is known if $L(\sigma)$ is given. Further differentiations yield

$$u_{xx}\,X'(\sigma) + u_{xy}\,Y'(\sigma) = M_1'(\sigma)$$

and $$u_{xy}\,X'(\sigma) + u_{yy}\,Y'(\sigma) = M_2'(\sigma) \text{ on } \Gamma \quad .$$

These equations, together with (11.6), define unique values of the second derivatives on Γ provided that

$$\Delta(\sigma) \equiv \begin{vmatrix} X'(\sigma) & Y'(\sigma) & 0 \\ 0 & X'(\sigma) & Y'(\sigma) \\ a(X,Y) & 2b(X,Y) & c(X,Y) \end{vmatrix} \neq 0 \quad . \tag{11.13}$$

Higher derivatives may similarly be calculated on Γ and, in principle, the resulting series solution may be used to determine Cauchy data on a neighbouring curve. Thus an integral surface may be generated by steps. However there is no guarantee of success in the passage to the limit, and for some systems the Cauchy data do not provide a satisfactory solution.

For a satisfactory solution we require three conditions:
(a) the solution exists, (b) the solution is unique, (c) the solution has a continuous dependence on the Cauchy data; the Cauchy problem is then said to be **well-posed.**

When $\Delta = 0$ the second derivatives of u are not uniquely determined on Γ and so the problem is not well posed. Expanding $\Delta = 0$ gives

$$a\,Y'^2(\sigma) - 2b\,Y'(\sigma)\,X'(\sigma) + cX'^2(\sigma) = 0 \quad .$$

If this is true for all values of σ, that is, at all points of Γ, then Γ is a characteristic. Consequently the problem is not well posed whenever Cauchy data are given on a characteristic. If $\Delta(\sigma) = 0$ at an isolated value of σ, then Γ touches a characteristic at some point T, and so the problem is well-posed only on part of the domain, for the Cauchy conditions must not be specified beyond the point of contact T, as in section 11.3.

11.5.2 The Riemann Solution

The canonical form of the hyperbolic equation is

$$\mathsf{L}(u) \equiv u_{\xi\eta} + \alpha(\xi,\eta)u_\xi + \beta(\xi,\eta)\,u_\eta + \gamma(\xi,\eta)\,u = F(\xi,\eta) \quad , \tag{11.14}$$

and the characteristics are the straight lines ξ constant, η constant. Corresponding to the operator $\mathbf{L}$ there is an **adjoint operator $\mathbf{M}$** defined by

$$\mathbf{M}(v) = v_{\xi\eta} - (\alpha v)_\xi - (\beta v)_\eta + \gamma v \quad .$$

Then
$$v\mathbf{L}(u) - u\mathbf{M}(v) = U_\xi + V_\eta \quad , \tag{11.15}$$

where $U = \frac{1}{2}(uv)_\eta - u(v_\eta - \alpha v)$ and $V = \frac{1}{2}(uv)_\xi - u(v_\xi - \beta v)$. Riemann's method of solving the Cauchy problem is to integrate (11.15) over a suitably chosen region $\mathfrak{D}$ with v taken as a specified solution of the adjoint homogeneous equation $\mathbf{M}(v) = 0$.

Let Cauchy data be given on a curve Γ which is nowhere tangential to a characteristic. Then the equation of Γ may be written as $\xi = \Xi(\eta)$ or $\eta = \mathrm{H}(\xi)$, where $\Xi(\eta)$ and $\mathrm{H}(\xi)$ are strictly monotone. Then if $P(\xi_0, \eta_0)$ is an arbitrary point, the characteristic lines $\xi = \xi_0$, $\eta = \eta_0$ meet Γ in $R[\xi_0, \mathrm{H}(\xi_0)]$ and $Q[\Xi(\eta_0), \eta_0]$ respectively (see Fig. 11.1).

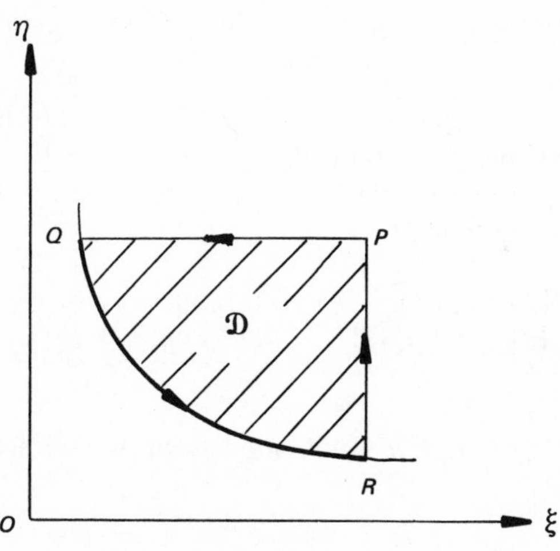

Fig. 11.1

Now let $\mathfrak{D}$ be the region bounded by the contour C consisting of the lines RP, PQ and the arc QR of Γ. Integrating (11.15) over $\mathfrak{D}$ and using (11.14) we have

$$\iint_{\mathfrak{D}} vF\,\mathrm{d}\xi\mathrm{d}\eta = \iint_{\mathfrak{D}} u\mathbf{M}(v)\mathrm{d}\xi\mathrm{d}\eta + \int_C (U\mathrm{d}\eta - V\mathrm{d}\xi) \quad . \tag{11.16}$$

Now
$$\int_{RP} (U\mathrm{d}\eta - V\mathrm{d}\xi) = \int_R^P U\mathrm{d}\eta = \tfrac{1}{2}(uv)_P - \tfrac{1}{2}(uv)_R - \int_R^P u(v_\eta - \alpha v)\mathrm{d}\eta$$

and $$\int_{PQ} (U\mathrm{d}\eta - V\mathrm{d}\xi) = \int_P^Q -V\mathrm{d}\xi = \tfrac{1}{2}(uv)_P - \tfrac{1}{2}(uv)_Q + \int_P^Q u(v_\xi - \beta v)\mathrm{d}\xi \,.$$

We now simplify the terms by choosing v to satisfy the following conditions:

$$\mathsf{M}(v) = 0 \text{ in } \mathfrak{D},$$
$$v_\eta = \alpha v \text{ on } \xi = \xi_0, \quad v_\xi = \beta v \text{ on } \eta = \eta_0 \;,$$
$$v(\xi_0, \eta_0) = 1 \;.$$

The function v is called the **Riemann function** of the Cauchy problem. It is usually easier to find v than to solve the Cauchy problem for u directly (for instance, for the classical wave equation in canonical form $u_{\xi\eta} = F$, the Riemann function $v = 1$).

With these conditions on v, (11.16) finally simplifies to

$$u(\xi_0, \eta_0) = \tfrac{1}{2}(uv)_Q + \tfrac{1}{2}(uv)_R + \int_{QR} (V\mathrm{d}\xi - U\mathrm{d}\eta) + \iint_{\mathfrak{D}} vF\mathrm{d}\xi\mathrm{d}\eta \quad . \text{ (11.17)}$$

This gives the required solution u in terms of the known functions v and F, and the known values (Cauchy data) of u, u_ξ, u_η on QR.

Finally we note that for the Riemann solution it is convenient to express the Cauchy data in the form

$$u[\xi, \mathrm{H}(\xi)] = a_1(\xi), \qquad u[\Xi(\eta),\eta] = a_2(\eta) \quad ,$$
$$u_\xi[\xi, \mathrm{H}(\xi)] = b_1(\xi), \qquad u_\eta[\Xi(\eta),\eta] = b_2(\eta) \quad .$$

Example 1

Solve $u_{\xi\eta} + (\xi + \eta)^{-1}(u_\xi + u_\eta) = 0$, with Cauchy data $u = 0$ and $u_\xi = f(\xi)$ on the line $\xi = \eta$.

Solution

The Riemann function satisfies

$$v_{\xi\eta} - \left(\frac{v}{\xi + \eta}\right)_\xi - \left(\frac{v}{\xi + \eta}\right)_\eta = 0$$

with $$v(\xi_0,\eta_0) = 1, v_\xi(\xi, \eta_0) = (\xi + \eta_0)^{-1} v(\xi,\eta_0)$$

and $$v_\eta(\xi_0,\eta) = (\xi_0 + \eta)^{-1} v(\xi_0,\eta) \quad .$$

Clearly $v(\xi, \eta) = (\xi + \eta)/(\xi_0 + \eta_0)$.

The region $\mathfrak{D}$ is the triangle with vertices $P(\xi_0, \eta_0)$, $Q(\eta_0, \eta_0)$, $R(\xi_0, \xi_0)$ and the Cauchy data is $a_1(\xi) = a_2(\eta) = 0$, $b_1(\xi) = f(\xi)$ and $b_2(\eta) = -f(\eta)$.

Then from (11.17)

$$u(\xi_0, \eta_0) = \tfrac{1}{2} \int_{QR} \left(\frac{\xi + \eta}{\xi_0 + \eta_0} f(\eta)\, \mathrm{d}\eta + \frac{\xi + \eta}{\xi_0 + \eta_0} f(\xi)\mathrm{d}\xi \right) .$$

As QR is the straight line $\xi = \eta$, we have, after dropping the suffices,

$$u(\xi, \eta) = \frac{2}{\xi + \eta} \int_{\eta}^{\xi} \xi' f(\xi')\mathrm{d}\xi' .$$

Example 2

For the equation $u_{\xi\eta} = F(\xi, \eta)$ the Riemann function $\nu(\xi, \eta) = 1$ and (11.17) gives

$$u(\xi, \eta) = \tfrac{1}{2} \left\{ a_1(\xi) + a_2(\eta) + \int_{\Xi(\eta)}^{\xi} b_1(\xi')\mathrm{d}\xi' + \int_{\mathrm{H}(\xi)}^{\eta} b_2(\eta')\mathrm{d}\eta' \right\}$$

$$+ \int_{\Xi(\eta)}^{\xi} \int_{\mathrm{H}(\xi')}^{\eta} F(\xi', \eta')\mathrm{d}\eta'\, \mathrm{d}\xi' .$$

11.5.3 Significance of the characteristics

The region $\mathfrak{D}$ is called the domain of dependence of the point P as the solution at P is determined solely by values of F in $\mathfrak{D}$ and Cauchy data on QR.

We also introduce the concept of the domain of influence of the arc QR of the curve Γ. This domain, denoted by $\mathfrak{R}$, is the set of all points P whose domains of dependence intersect QR. Thus solutions at points outside $\mathfrak{R}$ are not affected by the Cauchy data on QR. Clearly $\mathfrak{R}$ is the infinite region bounded by the arc QR and the characteristics ξ = constant and η = constant through Q and R respectively. (See Fig. 11.2.)

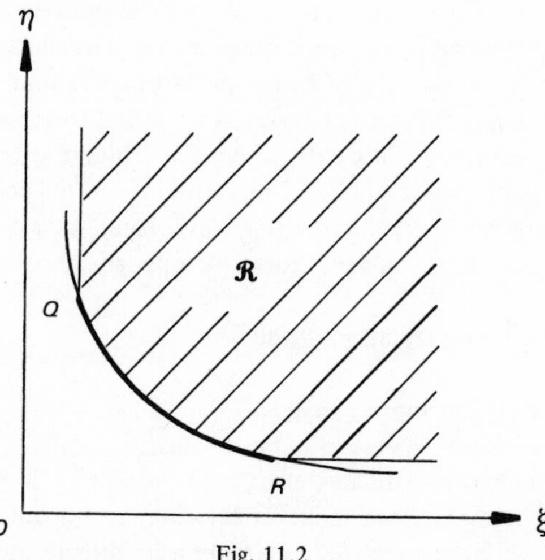

Fig. 11.2

Now consider any hyperbolic equation with two independent variables x and t, where x is a space variable and t is time, subject to **initial conditions**, that is, to Cauchy conditions on the line $t = 0$. The characteristics form a curvilinear coordinate system in xt-space and the points P, Q, R in the $\xi\eta$-plane correspond to points P_0, Q_0, R_0 in the xt-plane, where P_0 lies in the region $t > 0$ and Q_0, R_0 are on the x-axis. Fig. 11.3(a) shows $\mathfrak{D}_0$, the image of $\mathfrak{D}$ in xt-space. Thus $\mathfrak{D}_0$ is the curvilinear triangle $P_0Q_0R_0$ bounded by the line Q_0R_0 and the characteristics η = constant, ξ = constant passing through Q_0 and R_0 respectively. Similarly Fig. 11.3(b) shows $\mathfrak{R}_0$, the image of $\mathfrak{R}$, to be the infinite region bounded by the line Q_0R_0 and the characteristics ξ = constant, η = constant passing through Q_0 and R_0 respectively.

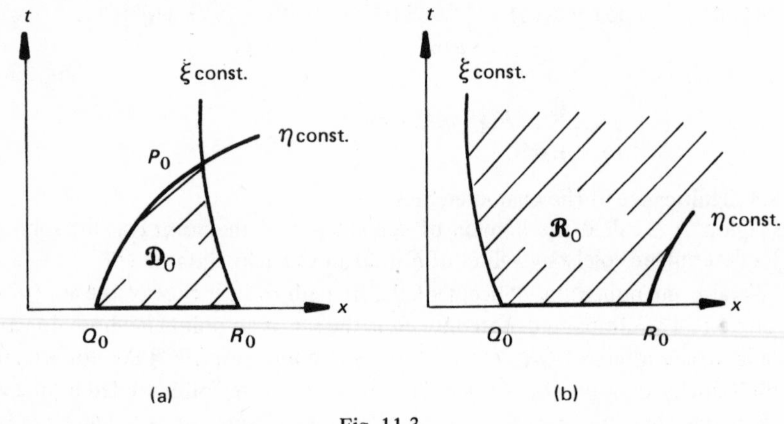

Fig. 11.3

The significance of the characteristics is that the boundaries of $\mathfrak{D}_0$ and $\mathfrak{R}_0$ consist of characteristics and an interval of the x-axis. The characteristics forming these boundaries can be thought of as the paths in space-time of signals which leave the end points Q_0, R_0 of the interval at $t = 0$ and travel with speeds c_1, c_2 in opposite directions along the x-axis, where the values of c_1 and c_2 correspond to the gradients of the characteristics. Thus when the initial conditions are non-zero only in the interval Q_0R_0 the resulting wave fronts, that is points where the disturbance first becomes non-zero, travel in opposite directions with speeds c_1 and c_2.

We deduce that for an equation of the form

$$c^2 u_{xx} = u_{tt} + \Phi(u, u_x, u_t) \quad ,$$

when the characteristic coordinates are $\xi = x - ct$, $\eta = x + ct$, and so the characteristics are straight lines in xt-space, the wave fronts from an initial disturbance travel with constant speed c, in opposite directions, along the x-axis.

PROBLEMS

11.1 Given that the curve (11.3) is a solution of (11.2), show that both ϕ and ψ satisfy

$$Pv_x + Qv_t + Rv_u = 0 \quad .$$

Deduce that

$$\frac{P}{J\begin{pmatrix}\phi & \psi\\ t & u\end{pmatrix}} = \frac{Q}{J\begin{pmatrix}\phi & \psi\\ u & x\end{pmatrix}} = \frac{R}{J\begin{pmatrix}\phi & \psi\\ x & t\end{pmatrix}} \quad .$$

Now consider the equation $F(\phi, \psi) = 0$, where F is arbitrary. Differentiate this equation with respect to both x and t, and, by eliminating F_ϕ, F_ψ, show that

$$u_x\, J\begin{pmatrix}\phi & \psi\\ t & u\end{pmatrix} + u_t\, J\begin{pmatrix}\phi & \psi\\ u & x\end{pmatrix} = J\begin{pmatrix}\phi & \psi\\ x & t\end{pmatrix} \quad .$$

Deduce that $F(\phi, \psi) = 0$ is a solution of (11.1).

11.2 Find the general solution of each of the following equations.
(a) $x^2u_x + t^2u_t = (x + t)u$
(b) $x(x + t)u_x - t(x + t)u_t = (t - x)(2x + 2t + u)$
(c) $(x + t)u_x + tu_t = u + t^2$
(d) $uxu_x - utu_t = t^2 - x^2$

11.3 For each of the following equations find the solution which satisfies the given Cauchy data.
(a) $2t(u - 3)u_x + (2x - u)u_t = t(2x - 3)$ when $u = 0$ on $x^2 - 2x + t^2 = 0$
(b) $(x - t)u_x + (t - x - u)u_t = u$ when $u = 1$ on $x^2 + t^2 = 1$
(c) $u_x + u_t = u$ when $u = \cos x$ on $t = 0$
(d) $x(t^2 + u)u_x - t(x^2 + u)u_t = (x^2 - t^2)u$ when $u = 1$ on $x + t = 0$
(e) $tu_x + xu_t = 1$ when $u = x$ on $x = 2t$

11.4 Reduce the following equations to canonical form.
(a) $y^2u_{xx} + x^2u_{yy} = 0$
(b) $x^2u_{xx} - y^2u_{yy} = 0$
(c) $x^2u_{xx} + 2xy\, u_{xy} + y^2u_{yy} = 0$
(d) $u_{xx} + 2u_{xy} + u_{yy} + u_x - u_y = 0$
(e) $u_{xx} + 2u_{xy} + 5u_{yy} + 3u_x + u = 0$
(f) $4u_{xx} - y^6u_{yy} - 3y^5u_y = 0$

11.5 Let $a_{11}u_{xx} + 2a_{12}u_{xt} + a_{22}u_{tt} + b_1u_x + b_2u_t + cu = 0$ be a linear second order equation with constant coefficients. Show that if the equation has a non-dispersive wave solution of the form $u(x, t) = f(x - At)$, where f is an arbitrary function and A is a constant, then

$$a_{11} - 2a_{12}A + a_{22}A^2 = 0 \quad , \quad b_1 - Ab_2 = 0, c = 0 \quad .$$

Deduce that

(a) equations of the elliptical type do not admit solutions in the form of waves with real velocities,

(b) for equations of the hyperbolic type two wave velocities of propagation are possible if $b_1 = b_2 = 0$,

(c) in the parabolic case there is one real velocity of propagation if $b_2 a_{12} = b_1 a_{22}$.

11.6 Solve the equation $u_{xx} = u_{tt}$ in the region $|x| < t < 2 - |x|$, given that $u = 0$ and $u_t = 1$ on the curve $2t = 1 + x^2$, $|x| < 1$.

11.7 Find a solution of the equation $u_{xx} = u_{tt} + xt$ satisfying the conditions $u(x, 0) = 0$ and $u_t(x, 0) = 0$.

11.8 Show that the equation $u_{xx} = u_{tt}$, with $u = a(x)$ and $u_t = b(x)$ on the line $x = t$, has no solution unless $a(x)$ and $b(x)$ satisfy the relation $2b(x) = a'(x) + k$ for some constant k. Show further that if $a(x)$, $b(x)$ satisfy such a relation then $u = f(x - t) + a(\frac{1}{2}x + \frac{1}{2}t) - f(0)$ is a solution for any suitably differentiable function $f(x)$, with $f'(0) = -\frac{1}{2}k$.

11.9 Use the Riemann method to find the solution of the equation

$\xi\eta u_{\xi\eta} - \xi u_\xi - \eta u_\eta + u = 0$ for which $u = \xi^3$ and $u_\xi = 0$ on $\xi = \eta$.

11.10 Use the Riemann method to derive the d'Alembert solution of the homogeneous wave equation.

11.11 Find the solution of the equation $x^2u_{xx} = t^2u_{tt}$ which satisfies the conditions $u(x, 1) = x^2$, $u_t(x, 1) = x^3$.

11.12 Prove that the Riemann function for the equation $4u_{\xi\eta} + u = 0$ is $v(\xi, \eta) = J_0(\sqrt{(\xi - \xi_0)(\eta - \eta_0)})$.
Hence determine the solution of the equation $u_{xx} + u = u_{tt}$ subject to the initial conditions $u(x, 0) = a(x)$, $u_t(x, 0) = b(x)$.

11.13 Prove that the equation $u_{\xi\eta} + 2(\xi + \eta)^{-1}(u_\xi + u_\eta) = 0$ has a Riemann function

$$\nu(\xi, \eta) = (\xi_0 + \eta_0)^{-3}(\xi + \eta)\,[2\xi\eta + (\xi_0 - \eta_0)(\xi - \eta) + 2\xi_0\eta_0]$$

Hence determine the solution which satisfies the Cauchy data $u = 0$, $u_\xi = 3\xi^2$ on $\xi = \eta$.

CHAPTER 12

The Cauchy Problem in Two and Three Dimensions

12.1 THE INITIAL VALUE PROBLEM IN THREE DIMENSIONS

We now consider the wave equation in space:

$$c^2\nabla^2 u = u_{tt},\ u = u(\mathbf{r}, t),\ \mathbf{r}\in\mathbb{R}^3,\ t > 0 \quad , \tag{12.1}$$

and seek a solution to it that satisfies the initial conditions

$$u(\mathbf{r}, 0) = a(\mathbf{r}),\ u_t(\mathbf{r}, 0) = b(\mathbf{r}) \quad . \tag{12.2}$$

To find the values $u(\mathbf{r}_0, t)$ of the solution at a point $P_0(\mathbf{r}_0)$ we introduce the spherically averaging operator $M(\mathbf{r}_0, R)$, which maps any field ϕ onto the mean value of ϕ on the sphere S with centre P_0 and radius R.
Thus

$$M(\mathbf{r}_0, R)[\phi] = \frac{1}{4\pi R^2}\int_S \phi(\mathbf{r})\ \mathrm{d}S \quad . \tag{12.3}$$

Now let $\quad M(\mathbf{r}_0, R)[u(\mathbf{r}, t)] = U(R, t) \quad , \qquad (12.4)$

$$M(\mathbf{r}_0, R)[a(\mathbf{r})] = A(R) \quad ,$$

and $\quad M(\mathbf{r}_0, R)[b(\mathbf{r})] = B(R) \quad .$

We note that $\quad U(0, t) = u(\mathbf{r}_0, t) \quad , \qquad (12.5)$

$$U(R, 0) = A(R) \quad , \tag{12.6}$$

and $\quad U_t(R, 0) = B(R) \quad . \qquad (12.7)$

We now also show that $U(R, t)$ satisfies the spherical wave equation

$$\frac{c^2}{R^2}\frac{\partial}{\partial R}\left(R^2\frac{\partial U}{\partial R}\right) = \frac{\partial^2 U}{\partial t^2} \quad . \tag{12.8}$$

From (12.3) and (12.4)

$$\frac{\partial U}{\partial R} = \frac{\partial}{\partial R}\int_S \frac{u(\mathbf{r}, t)}{4\pi}\frac{\mathrm{d}S}{R^2} \quad .$$

Now $\mathrm{d}S/R^2$ is the solid angle subtended at P_0 by the element $\mathrm{d}S$, that is, the projection of $\mathrm{d}S$ on to the concentric sphere S_1 of unit radius (see Fig. 12.1).

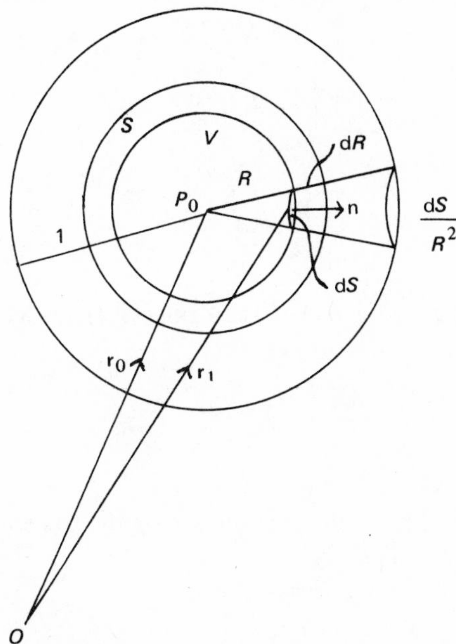

Fig. 12.1

Clearly this solid angle does not change as R increases. However, the values of u in the integrand increase with R at a rate of $\nabla u.\mathbf{n}$, where $\mathbf{n}$ is the unit outward normal. Hence

$$\frac{\partial U}{\partial R} = \int_S \frac{\nabla u.\mathrm{d}\mathbf{S}}{4\pi R^2} \quad .$$

It follows that

$$4\pi c^2 \frac{\partial}{\partial R}\left(R^2 \frac{\partial U}{\partial R}\right) = c^2 \frac{\partial}{\partial R}\int_S \nabla u.\mathrm{d}\mathbf{S}$$

$$= \frac{\partial}{\partial R}\int_V c^2 \nabla^2 u \,\mathrm{d}V$$

$$= \frac{\partial}{\partial R}\int_V u_{tt}(\mathbf{r}, t)\,\mathrm{d}V \quad ,$$

where V is the region enclosed by S. By considering the integral of u_{tt} over the spherical annulus of thickness dR shown in Fig. 12.1, we see that the rate of increase of the volume integral with R is just the integral of u_{tt} over the surface S. Hence

$$4\pi c^2 \frac{\partial}{\partial R} R^2\left(\frac{\partial U}{\partial R}\right) = \int_S u_{tt}(\mathbf{r}, t)\,\mathrm{d}S$$

$$= 4\pi R^2 \frac{\partial^2 U}{\partial t^2} \quad ,$$

and (12.8) is proved.

By substituting $U = R^{-1}v$ in (12.8) and using (2.2) it is seen that the general solution of (12.8) is

$$U = \frac{f(R - ct) + g(R + ct)}{R} \quad ,$$

where f and g are arbitrary C^2 functions defined on IR. To ensure the continuity of $U(R, t)$ at $R = 0$ we require

$$f(-ct) + g(ct) = 0, ct \in \mathrm{IR} \quad .$$

Hence $$U(R, t) = \frac{g(ct + R) - g(ct - R)}{R} \quad .$$

Now, from (12.5), (12.6) and (12.7) we obtain

$$u(\mathbf{r}_0, t) = U(0, t) = \lim_{R\to 0} U(R, t) = 2g'(ct) \quad ,$$

$$RA(R) = RU(R, 0) = g(R) - g(-R)$$

and

$$RB(R) = RU_t(R, 0) = cg'(R) - cg'(-R) \quad .$$

Hence $$2g'(R) = \frac{d}{dR}[RA(R)] + \frac{R}{c}B(R) \quad ,$$

which gives $u(\mathbf{r}_0, t) = \dfrac{1}{c}\dfrac{d}{dt}\,[ct\,A(ct)] + tB(ct) \quad .$

Therefore the required solution of (12.1) and (12.2) is

$$u(\mathbf{r}, t) = \frac{\partial}{\partial t}\left\{tM(\mathbf{r}, ct)[a]\right\} + tM(\mathbf{r}, ct)[b] \quad . \tag{12.9}$$

We see that the domain of dependence (as defined in Section 2.1) of the point P at time t is just the surface of the sphere with centre P and radius ct. This property of the domain of dependence being restricted to points at a distance ct from P holds for the wave equation in every space with an odd number n of dimensions except $n = 1$. It does not hold in spaces with an even number of dimensions.

The solution (12.9) was first given by Poisson, although the actual form (12.9) and the interpretation in terms of mean values over the sphere S are due to Stokes.

12.2 THE INITIAL VALUE PROBLEM IN TWO DIMENSIONS

When we introduce a Cartesian coordinate system (12.9) provides the solution of the homogeneous wave equation with initial conditions which are functions of the three coordinates x, y and z. If the initial functions a and b are independent of z, then clearly the solution u, given by (12.9), will also be independent of z. Thus the function $u = u(x, y, t)$ will satisfy

$$c^2(u_{xx} + u_{yy}) = u_{tt}$$

and the initial conditions

$$u(x, y, 0) = a(x, y), u_t(x, y, 0) = b(x, y) \quad .$$

So we may use the solution to the three-dimensional problem to solve the corresponding two-dimensional problem.

As the conditions and the solutions are the same for all points that are situated on a straight line parallel to the z-axis we need only consider the value of u at a point $P(x_0, y_0, 0) = P(x_0, y_0)$ on the xy-plane. Then, in (12.9), the integration over the sphere S can be expressed as an integration over the disk Σ, with centre P and radius ct, which is the intersection of the sphere S with the xy-plane (see Fig. 12.2). The element of surface $\mathrm{d}S$ has a projection $\mathrm{d}\sigma$ on the xy-plane, where

$$\mathrm{d}\sigma = \mathrm{d}S \cos\gamma$$

and

$$\cos\gamma = \frac{\sqrt{[(ct)^2 - (x_0 - x)^2 - (y_0 - y)^2]}}{ct} .$$

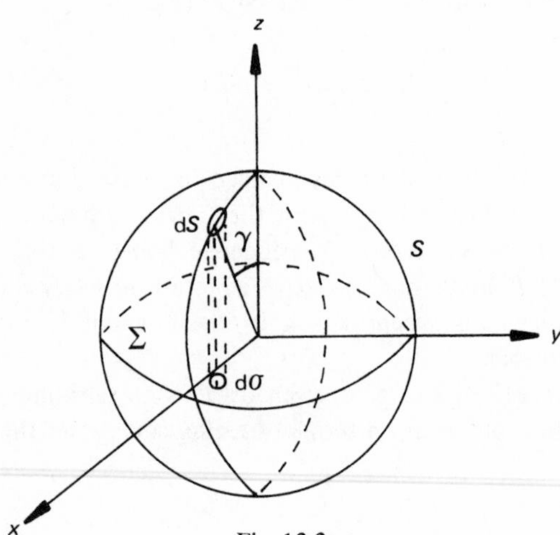

Fig. 12.2

As the same contribution occurs on integrating over the upper and lower half of the sphere the integral over the disk must be taken twice.

Thus (12.9) reduces to

$$u(x, y, t) = \frac{1}{2\pi c}\left\{\frac{\partial}{\partial t}\int_\Sigma \frac{a(x', y')\,\mathrm{d}x'\mathrm{d}y'}{\sqrt{[(ct)^2 - (x - x')^2 - (y - y')^2]}} + \int_\Sigma \frac{b(x', y')\,\mathrm{d}x'\mathrm{d}y'}{\sqrt{[(ct)^2 - (x - x')^2 - (y - y')^2]}}\right\} , \tag{12.10}$$

where the integration is taken over the interior of the circle of radius ct with centre at the point $P(x, y)$. Thus the domain of dependence of a point $P(x, y)$ consists of the interior and the boundary of a circle of radius ct and centre at P.

The procedure of using the solution of an equation in many variables to deduce the solution of the same problem for the equation with a smaller number of variables is called the **method of descent**. (see Problem 12.4 for descent from three dimensions to one).

12.3 HUYGHENS' PRINCIPLE

We now examine the solutions (12.9) and (12.10) together with the d'Alembert solution (2.9) of the one-dimensional problem. We shall consider, in the three cases, the propagation of a local disturbance, in which the initial functions a and b are non-zero only inside a bounded region G.

12.3.1 Propagation in three dimensions

To trace the sequence of events at a point P which lies outside the region G we construct a series of spheres with centre P and with continually increasing radii ct. Equation 12.9 shows that each such sphere is the domain of dependence of P at time t. For small values of t the spheres will not intersect G and so no effect is produced at P. If d_1, d_2 are the least and greatest distances from P to the boundary of G, the disturbance at P will begin after a time d_1/c, will last for a time $(d_2 - d_1)/c$, and will then return to zero. This phenomenon is known as **Huyghens' principle** for the wave equation. It states that an initial disturbance, localised in space, induces at every point of space an effect which is localised in time. (We encountered another form of Huyghens' principle, in the case of harmonic waves, in section 8.6.)

Now consider the three-dimensional region $H(t)$ which consists of points P where u is non-zero at time t. The points P are characterized by the fact that a sphere with centre P and radius ct intersects the region G. So to find the region H we construct at each point Q of G a sphere with centre Q and radius ct; this sphere is the region of influence of Q at time t. The envelope of the spheres will be the boundary of the region H (see Fig. 12.3). The outer surface is called the leading wave front and the inner surface the trailing wave front.

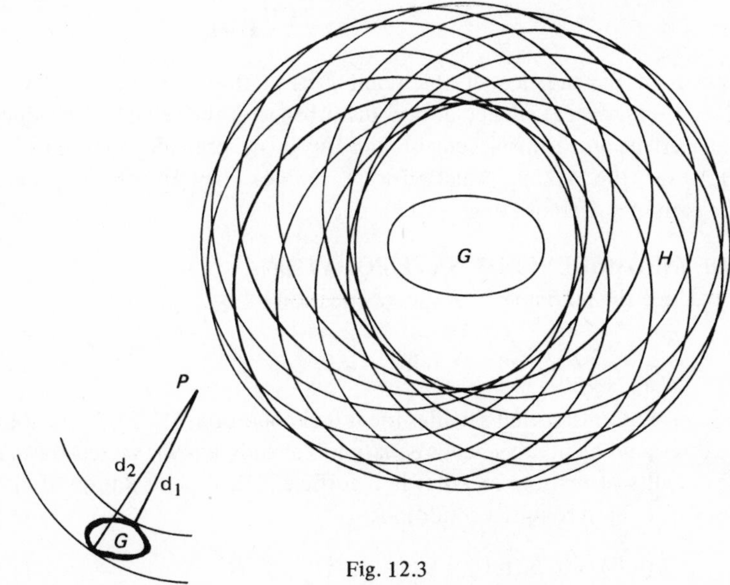

Fig. 12.3

Thus an initial disturbance in the region G causes a wave, with both its fronts, to pass through every point lying outside G. At the leading front the previously undisturbed points are excited, and at the trailing front the opposite effect occurs – the excited points return to an undisturbed state. The leading front reaches a given point P at time $t_1 = d_1/c$, and the trailing front leaves P at time $t_2 = d_2/c$.

12.3.2 Propagation in two dimensions

According to (12.10) the value of u at P at time t is determined by the initial values of u *throughout* the interior of the circle of radius ct with centre at P. Consequently for times $t < t_1 = d_1/c$ we again find that $u = 0$ at P, that is, the disturbance has not yet reached P. When $t = t_1$ a disturbance develops at P, but this disturbance does not cease at time $t_2 = d_2/c$, for the domain of dependence of P is now the whole disk, and after time t_2 the whole of G lies within it. It may be seen from (12.10) that what actually happens is that u decreases gradually and eventually tends to zero as $t \to \infty$. This phenomenon is called diffusion. The effect of initial disturbances, localised in a plane, is thus not localised in time and so Huyghens' principle does not hold.

We deduce that an instantaneous picture of the disturbances in a plane at any time t has a sharply defined leading front, but does not have a trailing front.

12.3.3 Propagation in one dimension

The d'Alembert solution (2.9) is

$$u(x, t) = \frac{a(x + ct) + a(x - ct)}{2} + \frac{1}{2c}\int_{x-ct}^{x+ct} b(x')\,\mathrm{d}x' . \tag{12.11}$$

The domain of dependence of the point $P(x)$ at time t is the whole of the interval $(x - ct, x + ct)$. Hence a local disturbance of the velocity b produces a permanent effect on the displacement, and Huyghens' principle does not strictly hold. However, the velocity function $u_t(x, t)$ does obey Huyghens' principle, as can be seen by differentiating (12.11).

12.4 THE INHOMOGENEOUS WAVE EQUATION

Let us examine the inhomogeneous wave equation

$$u_{tt} = c^2\nabla^2 u + p(\mathbf{r}, t), \mathbf{r}\in\mathbb{R}^3, t > 0 \tag{12.12}$$

and seek a solution to it that satisfies the initial conditions (12.2). As the solution of the corresponding homogeneous equation is already known we see, by analogy with the results of section 2.9, that it is sufficient to find a solution of (12.12) which satisfies the zero initial conditions

$$u(\mathbf{r}, 0) = 0, u_t(\mathbf{r}, 0) = 0 \ . \tag{12.13}$$

To find this solution we use the following generalised form of Duhamel's principle (see sections 1.3.3 and 2.9):

Define $\psi(\mathbf{r}, t; \tau)$, a function of position $\mathbf{r}$, time t and a parameter τ, as the solution of

$$\psi_{tt} = c^2\nabla^2\psi$$

satisfying

$$\psi(\mathbf{r}, 0; \tau) = 0, \psi_t(\mathbf{r}, 0; \tau) = p(\mathbf{r}, \tau) \quad .$$

Then

$$W(\mathbf{r}, t) = \int_0^t \psi(\mathbf{r}, t - \tau; \tau)\mathrm{d}\tau$$

is a solution of equation (12.12) with the initial conditions (12.13). This can be proved by substitution, as in section 1.3.3.

From (12.9) we have

$$\psi(\mathbf{r}, t - \tau; \tau) = (t - \tau)M(\mathbf{r}, c(t - \tau))[p(\mathbf{r}, \tau)] \quad .$$

Hence, writing $t - \tau = \sigma$, we obtain

$$W(\mathbf{r}, t) = \int_0^t \sigma M(\mathbf{r}, c\sigma)[p(\mathbf{r}, t - \sigma)] \, \mathrm{d}\sigma \quad . \tag{12.14}$$

The expression shown in (12.14) is called the **retarded potential**. It should be noted that, in the integration in (12.14), the function p is taken at the instant of time t being considered but at the **retarded time** $t' = t - \sigma$. The difference, $t - t' = \sigma$, is the time taken for the process being propagated with speed c to travel from the centre to the surface of a sphere of radius $c\tau$.

The same procedure may be used to obtain a solution to the two-dimensional inhomogeneous wave equation

$$u_{tt} = c^2(u_{xx} + u_{yy}) + p(x, y, t)$$

with initial conditions

$$u(x, y, 0) = 0, u_t(x, y, 0) = 0 \quad .$$

In this case the function ψ is constructed from (12.10), and the required solution is the function

$$W(x, y, t) = \frac{1}{2\pi c}\int_0^t \int_\Sigma \frac{p(x', y', \tau)}{\sqrt{[c^2(t - \tau)^2 - (x - x')^2 - (y - y')^2]}} \mathrm{d}x'\mathrm{d}y'\mathrm{d}\tau, \tag{12.15}$$

where Σ is the interior of the circle of radius $c(t-\tau)$ with centre at the point $P(x, y)$.

12.5 A UNIQUENESS THEOREM FOR THE CAUCHY PROBLEM

The solution of the Cauchy problem for the wave equation is unique. For greater clarity we consider only the case of two dimensions, but the method can be extended to higher dimensions (see Courant and Hilbert 1962, Ch. VI).

To prove that the solution of

$$u_{tt} = c^2(u_{xx} + u_{yy}) \tag{12.16}$$

subject to the initial conditions

$$u(x, y, 0) = a(x, y), u_t(x, y, 0) = b(x, y) \tag{12.17}$$

is unique we need to show that $u = 0$ is the only solution of (12.16) subject to the zero initial conditions

$$u(x, y, 0) = 0, u_t(x, y, 0) = 0 \tag{12.18}$$

Consider an arbitrary point $P(x_0, y_0, t_0)$, with $t_0 > 0$, in xyt-space and construct the cone

$$(x - x_0)^2 + (y - y_0)^2 = c^2(t - t_0)^2$$

with the point P as vertex (see Fig. 12.4). Let V be the region bounded by the surface Γ of the cone and the two disks S_0, S_1 bounded by its intersections with the planes $t = 0$ and $t = t_1$, where $0 < t_1 < t_0$. It is easily verified that

$$2u_t(c^{-2}u_{tt} - u_{xx} - u_{yy}) = (u_x^2 + u_y^2 + c^{-2}u_t^2)_t - 2(u_tu_x)_x - 2(u_tu_y)_y$$

and when this identity is integrated over the region V the integral on the left hand side vanishes, since u is a solution of (12.16). Hence, using the divergence theorem, we obtain

$$\iint\limits_{\Gamma+S_0+S_1} \left[(u_x^2 + u_y^2 + \frac{1}{c^2}u_t^2)l - 2u_tu_xm - 2u_tu_yn\right] dS = 0 \tag{12.19}$$

where the integration is over the surface of V and l, m, n are the direction cosines of the normal relative to the Ot, Ox and Oy axes. On S_1, $l = 1$, $m = n = 0$ and so the integral over S_1 is

$$\iint\limits_{S_1} (u_x^2 + u_y^2 + \frac{1}{c^2}u_t^2)dS \quad . \tag{12.20}$$

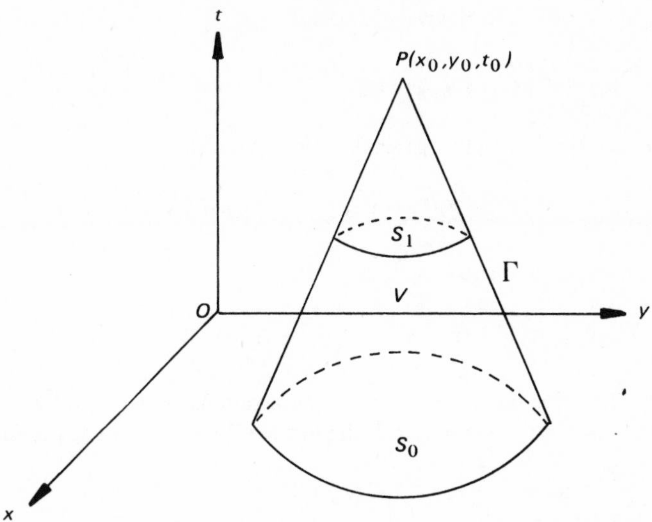

Fig. 12.4

The initial conditions (12.18) imply that u and all its first derivatives are zero on S_0. Hence the integral over S_0 vanishes. On Γ, $l^2 = m^2 + n^2$, and so the integral over Γ may be written

$$\iint_\Gamma \frac{1}{l}\left[(u_x l - u_t m)^2 + (u_y l - u_t n)^2\right] \mathrm{d}S \geqslant 0 \quad ,$$

because $l > 0$ on Γ. Hence we deduce that the integral (12.20) must vanish, and so the first order partial derivatives of u are equal to zero at all points within the entire cone. Thus u is constant. However $u = 0$ on the base S_0 and so we conclude that $u = 0$ at the point P. Consequently the solution of (12.16) and (12.17) is unique.

PROBLEMS

12.1 Find the solution to the wave equation $u_{tt} = c^2\nabla^2 u$ that satisfies the initial conditions

(a) $u(\mathbf{r}, 0) = 0, u_t(\mathbf{r}, 0) = \begin{cases} \alpha \text{ within a sphere of radius } R \\ 0 \text{ outside this sphere} \end{cases}$,

(b) $u(\mathbf{r}, 0) = \begin{cases} \beta \text{ within a sphere of radius } R \\ 0 \text{ outside this sphere} \end{cases}$, $u_t(\mathbf{r}, 0) = 0$.

12.2 Find the solution to the wave equation

$$u_{tt} = c^2(u_{xx} + u_{yy} + u_{zz})$$

in the half-space $z > 0$ that satisfies the initial conditions

$$u(x, y, z, 0) = a(x, y, z), u_t(x, y, z, 0) = b(x, y, z)$$

and the boundary conditions

$$u(x, y, 0, t) = 0 \quad \text{or} \quad u_z(x, y, 0, t) = 0 \quad .$$

12.3 Solve the wave equation $u_{tt} = c^2\nabla^2 u$ assuming that the initial value of u_t is everywhere equal to zero and that the initial value of $u = a$ has the form

$$a = \begin{cases} \alpha \cos \dfrac{\pi}{2R} r \;, & r \leqslant R \\ 0 \;, & r > R \end{cases} \quad .$$

12.4 Obtain the d'Alembert solution (2.9) to the initial value problem for the equation $c^2 u_{xx} = u_{tt}$ by the method of descent of section 12.2. The integration over S can be expressed as an integration over the interval $(x_0 - ct, x_0 + ct)$ of the x-axis.

CHAPTER 13

Appendix: Special Functions

13.1 THE GAMMA FUNCTION

The gamma function $\Gamma(x)$ is defined for $x > 0$ by

$$\Gamma(x) = \int_0^\infty e^{-t} t^{x-1} \, dt \quad . \tag{13.1}$$

Integrating (13.1) by parts gives

$$\Gamma(x) = x^{-1} \Gamma(x+1) \quad . \tag{13.2}$$

Now $\Gamma(1) = 1$, and so

$$\Gamma(x+1) = x! \, , \quad (x = 0, 1, 2, \ldots) \quad . \tag{13.3}$$

Since $\Gamma(x + 1)$ is well defined for $x > -1$, (13.2) may be used to extend the definition of $\Gamma(x)$ to the domain $-1 < x < 0$. Then, by repeated application of (13.2), $\Gamma(x)$ can be defined for all real values of x except at the points $0, -1, -2, \ldots$, at which $\Gamma(x)$ becomes infinite.

13.2 BESSEL FUNCTIONS

13.2.1 Bessel functions of the first kind

Bessel's equation of order n is

$$x^2 \frac{d^2 y}{dx^2} + x \frac{dy}{dx} + (x^2 - n^2) y = 0 \quad , \tag{13.4}$$

and since it is only n^2 that occurs in the equation, we may assume that $n \geqslant 0$.

As (13.4) has a regular singularity at $x = 0$ we may use the **method of Frobenius** to find a solution of the form

$$y(x) = x^p \sum_{r=0}^{\infty} a_r x^r , \qquad a_0 \neq 0 \quad .$$

This leads to the indicial equation

$$p^2 - n^2 = 0$$

together with the equations

$$[(p+1)^2 - n^2]\, a_1 = 0$$

and $$[(p+r)^2 - n^2]\, a_r + a_{r-2} = 0, \qquad r \geqslant 2 \quad .$$

Consequently $p = \pm n, \quad a_1 = 0, \quad a_{2r+1} = 0 \quad ,$

and we find that $a_r = (-1)^n \dfrac{a_0}{2^{2r} r!(p+1)(p+2)\ldots(p+r)}, \qquad (p = \pm n) \quad ,$

which may be written

$$a_{2r} = (-1)^n a_0 \frac{\Gamma(p+1)}{2^{2r} r!\, \Gamma(p+r+1)} \quad .$$

As a_0 is an arbitrary constant we may set $a_0 = 1/[2^p\, \Gamma(p+1)]$ and obtain two solutions of (13.4), denoted by $J_n(x)$ and $J_{-n}(x)$, where

$$J_p(x) = \sum_{r=0}^{\infty} (-1)^r \frac{1}{r!\Gamma(p+r+1)} \left(\frac{x}{2}\right)^{2r+p} \quad . \tag{13.5}$$

The functions $J_n(x)$ and $J_{-n}(x)$ are called **Bessel functions of the first kind.**

When x is small the series representation of $J_p(x)$ is dominated by the term $(x/2)^p/\Gamma(p+1)$ and so, provided n is not integral, we see that $J_n(x)$ and $J_{-n}(x)$ and independent solutions of (13.4).

13.2.2 Bessel functions of the second kind

When n is an integer $\Gamma(-n+r+1) = \infty$ for all $r = 0, 1, 2, \ldots (n-1)$, and so we deduce from (13.5) that

$$J_{-n}(x) = (-1)^n J_n(x), \qquad (n = 0, 1, 2, \ldots) \quad . \tag{13.6}$$

Thus $J_n(x)$ and $J_{-n}(x)$ are not independent solutions. A series solution independent of $J_n(x)$ may be found by using the standard Frobenius procedure for the special case when the roots of the indicial equation differ by an integer.

However, it is more convenient to use a second solution which can be represented by the same formula for both integer and non-integer n.

We define the **Bessel function of the second kind** $Y_n(x)$ as

$$Y_n(x) = \frac{\cos n\pi \, J_n(x) - J_{-n}(x)}{\sin n\pi} \quad , \quad (n \text{ not an integer}) \quad . \tag{13.7}$$

When n is not an integer $Y_n(x)$ is a linear combination of $J_n(x)$ and $J_{-n}(x)$ and as such is a solution which is independent of $J_n(x)$.

For integer n (13.7) has the form 0/0. We extend the definition to this case by defining

$$Y_n(x) = \lim_{m \to n} \frac{\cos m\pi \, J_m(x) - J_{-m}(x)}{\sin m\pi} \quad . \tag{13.8}$$

This formula clearly defines $Y_n(x)$ for *all* values of n. Then using L'Hôpital's rule and (13.6), we find that, for n an integer,

$$Y_n(x) = \frac{1}{\pi}\left[\frac{\partial}{\partial n} J_n(x) - (-1)^n \frac{\partial}{\partial n} J_{-n}(x)\right] \quad . \tag{13.9}$$

It is now necessary to prove that (13.9) is a solution of (13.4) and also that it is independent of $J_n(x)$. The first of these can be proved by substituting (13.9) for y in (13.4), observing that both $J_n(x)$ and $J_{-n}(x)$ are solutions of (13.4), and using (13.6).

The independence of $Y_n(x)$ and $J_n(x)$ for integer n is proved by examining the behaviour of the functions when x is small (section 13.2.4).

13.2.3 Recurrence relations for Bessel functions

If $C_n(x) = A\,J_n(x) + B\,Y_n(x)$, where A and B are constants, then

$$(x^n C_n)' = x^n C_{n-1} \text{ and } (x^{-n} C_n)' = -x^{-n} C_{n+1} \quad . \tag{13.10}$$

These relations are easily shown to hold for $J_n(x)$ by using the series (13.5), and hence also for $Y_n(x)$ as defined by (13.8).

It follows that

$$C_n{}' = C_{n-1} - \frac{n}{x} C_n \text{ and } C_n{}' = \frac{n}{x} C_n - C_{n+1} \quad . \tag{13.11}$$

13.2.4 The behaviour of $J_n(x)$ and $Y_n(x)$ for small values of x

THEOREM 13.1

As $x \to 0$

$$J_n(x) \sim \frac{1}{\Gamma(n+1)} \left(\frac{x}{2}\right)^n , \tag{13.12}$$

$$Y_n(x) \sim -\frac{1}{\pi} \Gamma(n) \left(\frac{2}{x}\right)^n \quad (n > 0) , \tag{13.13}$$

$$Y_0(x) \sim \frac{2}{\pi} \ln x . \tag{13.14}$$

Proof

(i) Formula (13.12) is simply the term with the lowest power of x in the series (13.5).

(ii) When n is not an integer the dominant term in (13.7) is the first term of the series for $-J_{-n}(x)$ cosec $n\pi$, which, from (13.5), is

$$-\frac{1}{\sin n\pi} \frac{1}{\Gamma(1-n)} \left(\frac{2}{x}\right)^n .$$

Using the known property of the gamma function that

$$\Gamma(n)\,\Gamma(1-n) \sin n\pi = \pi ,$$

we obtain

$$Y_n(x) \sim -\frac{1}{\pi} \Gamma(n) \left(\frac{2}{x}\right)^n .$$

This is a continuous function of n, and therefore the formula survives the limiting proces in (13.8) and remains valid when n is a positive integer.

(iii) From (13.9), $Y_0(x) = \frac{1}{\pi} \lim\limits_{n \to 0} \left(\frac{\partial J_n}{\partial n} - \frac{\partial J_{-n}}{\partial n}\right)$.

The result (13.14) is derived by substituting for J_n and J_{-n} using the leading term of (13.5).

We note that $Y_n(0) = -\infty$ for all $n \geqslant 0$, whilst $J_0(0) = 1$, and $J_n(0) = 0$ when $n > 0$.

13.2.5 The behaviour of $J_n(x)$ and $Y_n(x)$ for large values of x.

When we set $y(x) = Z(x)/\sqrt{x}$ Bessel's equation (13.4) transforms into the equation

$$\frac{d^2Z}{dx^2} + \left(1 + \frac{1-4n^2}{4x^2}\right) Z = 0 . \tag{13.15}$$

Hence for large values of x both $J_n(x)$ and $Y_n(x)$ have an asymptotic form $(A \cos x + B \sin x)/\sqrt{x}$, for some values of the constants A and B. Determining the particular values of A and B that correspond to $J_n(x)$ and $Y_n(x)$ involves long and complex analysis. We simply note that as $x \to \infty$

$$J_n(x) \sim \left(\frac{2}{\pi x}\right)^{1/2} \cos\left[x - \frac{(2n+1)\pi}{4}\right] \tag{13.16}$$

and

$$Y_n(x) \sim \left(\frac{2}{\pi x}\right)^{1/2} \sin\left[x - \frac{(2n+1)\pi}{4}\right] . \tag{13.17}$$

13.2.6 Fourier-Bessel series

THEOREM 13.2

If $f(x)$ is defined in the interval $0 \leqslant x \leqslant a$ and can be expanded in the form

$$f(x) = \sum_{r=1}^{\infty} C_r J_n(\zeta_{nr} x/a) \quad ,$$

where ζ_{nr} are the roots of the equation $J_n(\zeta) = 0$, then the constants C_r are given by

$$C_r a^2 J_{n+1}^2(\zeta_{nr}) = 2 \int_0^a x f(x) J_n(\zeta_{nr} x/a) \mathrm{d}x \quad . \tag{13.18}$$

The proof of this theorem is an immediate consequence of the following theorem.

THEOREM 13.3

If $J_n(\zeta_{nr}) = 0$ and $J_n(\zeta_{ns}) = 0$,

then

$$\int_0^1 x J_n(\zeta_{nr} x) J_n(\zeta_{ns} x) \mathrm{d}x = \tfrac{1}{2} J_{n+1}^2(\zeta_{nr}) \delta_{rs} \quad , \tag{13.19}$$

where δ_{rs} is the Kronecker delta.

Proof

We first show that if $\zeta_{nr} \neq \zeta_{ns}$ then

$$\int_0^1 x J_n(\zeta_{nr} x) J_n(\zeta_{ns} x) \, \mathrm{d}x = 0 \quad .$$

From (13.4) we see that $J_n(\zeta_{nr} x)$ and $J_n(\zeta_{ns} x)$ satisfy the equations

$$x^2 \frac{\mathrm{d}^2}{\mathrm{d}x^2} J_n(\zeta_{nr} x) + x \frac{\mathrm{d}}{\mathrm{d}x} J_n(\zeta_{nr} x) + (\zeta_{nr}^2 x^2 - n^2) J_n(\zeta_{nr} x) = 0 \tag{13.20}$$

and

$$x^2 \frac{d^2}{dx^2} J_n(\zeta_{ns}x) + x \frac{d}{dx} J_n(\zeta_{ns}x) + (\zeta_{ns}^2 x^2 - n^2) J_n(\zeta_{ns}x) = 0 \quad . \tag{13.21}$$

Multiplying (13.20) and (13.21) by $J_n(\zeta_{ns}x)$ and $J_n(\zeta_{nr}x)$ respectively and subtracting gives, after some rearrangement,

$$\frac{d}{dx}\left[xJ_n(\zeta_{nr}x) \frac{d}{dx} J_n(\zeta_{ns}x) - x\,J_n(\zeta_{ns}x) \frac{d}{dx} J_n(\zeta_{nr}x)\right]$$
$$+ (\zeta_{nr}^2 - \zeta_{ns}^2)\, x\, J_n(\zeta_{nr}x) J_n(\zeta_{ns}x) = 0 \quad .$$

Integrating from 0 to 1 and using

$$J_n(\zeta_{nr}) = J_n(\zeta_{ns}) = 0 \text{ and } \zeta_{nr} \neq \zeta_{ns}$$

gives
$$\int_0^1 x\, J_n(\zeta_{nr}x) J_n(\zeta_{ns}x)\, dx = 0 \quad .$$

To complete the proof we need to show that

$$\int_0^1 x\, J_n^2(\zeta x)\, dx = \tfrac{1}{2} J_{n+1}^2(\zeta) \tag{13.22}$$

whenever $J_n(\zeta) = 0$.

If in (13.20) we replace ζ_{nr} by ζ and multiply throughout by $2(d/dx) J_n(\zeta x)$ the resulting equation is readily seen to be equivalent to

$$\frac{d}{dx}\left[x^2 \left(\frac{d}{dx} J_n(\zeta x)\right)^2 - n^2 J_n^2(\zeta x) + \zeta^2 x^2 J_n^2(\zeta x)\right] - 2\zeta^2 x\, J_n^2(\zeta x) = 0 \quad .$$

Hence, integrating from 0 to 1 and using the facts that $J_n(\zeta) = 0$ and $n J_n(0) = 0$,

$$2\zeta^2 \int_0^1 x\, J_n^2(\zeta x)\, dx = \left(\zeta\, J_n{}'(\zeta)\right)^2 \quad .$$

Now, using (13.11) we have $J_n{}'(\zeta) = -J_{n+1}(\zeta)$, and so (13.22) is proved.

13.3 LEGENDRE FUNCTIONS

13.3.1 Legendre polynomials

Legendre's differential equation is

$$(1 - x^2) \frac{d^2 y}{dx^2} - 2x \frac{dy}{dx} + n(n+1) y = 0 \quad . \tag{13.23}$$

As $x = 0$ is a regular point we may obtain a Taylor series expansion for $y(x)$ in the form

$$y(x) = \sum_{r=0}^{\infty} a_r \frac{x^r}{r!}, \quad \text{where } a_r = \frac{d^r y(0)}{dx^r} \quad .$$

Differentiating (13.23) r times gives

$$a_{r+2} = -(n-r)(n+r+1)\,a_r \quad ,$$

and so the general solution is

$$y(x) = A\,u(x) + B\,v(x) \quad , \tag{13.24}$$

where $\quad u(x) = 1 - n(n+1)\dfrac{x^2}{2!} + n(n+1)(n-2)(n+3)\dfrac{x^4}{4!} + \ldots,$

$$v(x) = x - (n-1)(n+2)\frac{x^3}{3!} + (n-1)(n+2)(n-3)(n+4)\frac{x^5}{5!} + \ldots$$

and A, B are arbitrary constants.

Standard convergence tests show that, if u and v are infinite series they converge for $|x| < 1$ and diverge for $|x| = 1$; moreover, (13.24) always diverges either at $x = 1$ or at $x = -1$. However in many physical applications solutions to (13.23) are required which are finite for $|x| = 1$. Such solutions exist when n is an integer, for then, if $n = 0, 2, 4 \ldots, u(x)$ is a polynomial, and if $n = 1, 3, 5 \ldots, v(x)$ is a polynomial. It is clear that these are the only solutions finite at $|x| = 1$. No new solutions arise from negative integer values of n, because, putting $n = -m$ in (13.23), where $m \geqslant 1$, we have $n(n+1) = (m-1)m$. We define the **Legendre polynomial** $P_n(x)$ as the polynomial solution of degree n normalised by the condition $P_n(1) = 1$. By considerable manipulation of the expressions for $u(x)$ and $v(x)$ it is possible to show that

$$P_n(x) = \frac{1}{2^n} \sum_{r \leqslant n/2} \frac{(-1)^r (2n-2r)!}{(n-r)!\, r!\, (n-2r)!} x^{n-2r}, \; (n = 0, 1, 2, \ldots) \quad . \tag{13.25}$$

Note that $P_n(x)$ is an even function if n is even, and an odd function if n is odd. In particular, $P_n(-1) = (-1)^n$. Also we may write

$$P_n(x) = \frac{1}{2^n} \sum_r \frac{(-1)^r}{(n-r)!\, r!} \frac{d^n}{dx^n} (x^{2n-2r})$$

$$= \frac{1}{2^n n!} \frac{d^n}{dx^n} \left[\sum_r \frac{(-1)^r n!}{r!(n-r)!} x^{2n-2r} \right] \quad ,$$

from which follows Rodrigues' formula

$$P_n(x) = \frac{1}{2^n n!} \frac{\mathrm{d}^n}{\mathrm{d}x^n} (x^2 - 1)^n \quad . \tag{13.26}$$

The first few polynomials are

$$P_0(x) = 1; \; P_1(x) = x \; ; \; P_2(x) = \tfrac{1}{2}(3x^2 - 1);$$

$$P_3(x) = \tfrac{1}{2}(5x^3 - 3x); \; P_4(x) = \tfrac{1}{8}(35x^4 - 30x^2 + 3) \quad .$$

The generating function $F(t, x)$ for Legendre polynomials is given by

$$F(t, x) = \frac{1}{\sqrt{(1 - 2tx + t^2)}} = \sum_{n=0}^{\infty} t^n P_n(x), \quad |t| < 1, |x| \leqslant 1 \quad . \tag{13.27}$$

This result may be proved by using the Binomial theorem to expand the left hand side, together with formula (13.25).

13.3.2 Recurrence relations

Differentiating (13.27) with respect to t gives

$$(1 - 2tx + t^2) \sum_{n=0}^{\infty} nt^{n-1} P_n(x) = (x - t) \sum_{n=0}^{\infty} t^n P_n(x) \quad .$$

Now equating coefficients of t^n we obtain

$$(n + 1) P_{n+1} + n P_{n-1} = (2n + 1) x P_n \quad . \tag{13.28}$$

Differentiating (13.27) with respect to x gives

$$(1 - 2tx + t^2) \sum_{n=0}^{\infty} t^n P_n'(x) = t \sum_{n=0}^{\infty} t^n P_n(x) \quad .$$

Again equating coefficients of t^n we obtain

$$P_n' - 2x P_{n-1}' + P_{n-2}' = P_{n-1} \quad . \tag{13.29}$$

13.3.3 Fourier-Legendre series

If $f(x)$ is defined in the interval $|x| \leqslant 1$ and can be expanded in the form

$$f(x) = \sum_{n=0}^{\infty} c_n P_n(x) \quad ,$$

then

$$c_n = (n + \tfrac{1}{2}) \int_{-1}^{1} f(x) P_n(x)\,\mathrm{d}x \quad . \tag{13.30}$$

This result follows from the relation

$$\int_{-1}^{1} P_m(x) P_n(x)\,\mathrm{d}x = \frac{2}{2n+1}\,\delta_{mn} \quad , \tag{13.31}$$

which we now prove.

Proof

Legendre's equation (13.23) may be written in the form

$$\frac{\mathrm{d}}{\mathrm{d}x}\left[(1-x^2)\frac{\mathrm{d}y}{\mathrm{d}x}\right] + n(n+1)y = 0 \quad .$$

Then the orthogonality expressed by (13.31) when $m \neq n$ is an immediate consequence of the orthogonality of the eigenfunctions of a Sturm-Liouville system (see section 4.2.4).

It remains to prove that

$$\int_{-1}^{1} P_n^2(x)\,\mathrm{d}x = \frac{2}{2n+1} \quad . \tag{13.32}$$

From the generating function (13.27) we have

$$\begin{aligned}
\frac{1}{1-2tx+t^2} &= \left[\sum_{n=0}^{\infty} t^n P_n(x)\right]^2 \\
&= \left[\sum_{m=0}^{\infty} t^m P_m(x)\right]\left[\sum_{n=0}^{\infty} t^n P_n(x)\right] \\
&= \sum_{m=0}^{\infty} \sum_{n=0}^{\infty} t^{m+n} P_m(x) P_n(x) \quad .
\end{aligned}$$

Integrating both sides with respect to x between -1 and 1 and using the orthogonality result gives

$$\sum_{n=0}^{\infty} t^{2n} \int_{-1}^{1} P_n^2(x)\,\mathrm{d}x = \frac{1}{t}\ln\frac{1+t}{1-t} = 2\sum_{n=0}^{\infty} \frac{t^{2n}}{2n+1} \quad .$$

Equating coefficients of powers of t then gives (13.32).

Answers to Problems and Hints

ANSWERS TO PROBLEMS AND HINTS

1.1 (iv) Write $\omega t = \alpha + \phi$ and $\mathbf{C}e^{-i\alpha} = \mathbf{A}(\alpha) + i\mathbf{B}(\alpha)$. Then, if $\mathbf{C}.\mathbf{C}e^{-2i\alpha}$ is real, $\mathbf{A}(\alpha) \perp \mathbf{B}(\alpha)$ and the equation $\mathbf{u} = \mathbf{A}(\alpha)\cos\phi + \mathbf{B}(\alpha)\sin\phi$ represents an ellipse with principal axes parallel to $\mathbf{A}(\alpha)$ and $\mathbf{B}(\alpha)$. (v) Hence the extreme values of OP are $|\mathbf{A}(\alpha)| = \frac{1}{2}|\mathbf{C}e^{-i\alpha} + \bar{\mathbf{C}}e^{i\alpha}|$ and $|\mathbf{B}(\alpha)| = \frac{1}{2}|\mathbf{C}e^{-i\alpha} - \bar{\mathbf{C}}e^{i\alpha}|$.

1.4 From $A_m = A\sqrt{2}$ we find $|p - p_m| = 2\lambda|p + p_m|^{-1}\sqrt{(\beta^2 - \lambda^2)}$. Approximating $p \cong p_m \cong \beta$, the two roots satisfy $|p_1 - p_2| \cong 2\lambda$.

1.5 In order of increasing λ the results for $t = 5$ are (.08, .04, .17); $t = 10$ gives (.007, .0005, .026); $t = 20$ gives $(5 \times 10^{-5}, 4 \times 10^{-8}, 6 \times 10^{-4})$.

1.7 $$w = \frac{ie^{-2t}}{6}\left(\frac{e^{3it}}{2-5i} - \frac{e^{-3it}}{2+i}\right) + \frac{e^{-2it}}{9-8i}\ .$$

1.8 $b = \frac{1}{2}ic\beta^{-1}$. Phase retarded by $\frac{1}{2}\pi$.

1.11 $Ae^{ikx_R} + Be^{-ikx_R} = 0$. Then
$Z_P(\omega) = V(x_P)/I(x_P) = -i\omega Lk^{-1}\tan kh = -iZ\tan kh$.

2.4 The region $t \geqslant 0, x + ct \geqslant x_1, x - ct \leqslant x_2$.

2.5 (i) $u = 0\ (x \geqslant ct)$, $\frac{1}{2}a(nc)^{-1}[\cos n(x - ct) - 1]\ (-ct \leqslant x \leqslant ct)$, $a(nc)^{-1}\sin nx\ \sin nct\ (x \leqslant -ct)$
(ii) $u = \frac{1}{2}h\{[1 + (x + ct)^2]^{-1} + [1 + (x - ct)^2]^{-1}\}$.
(iii) $u = 0\ (x > \alpha + ct \text{ or } x < -ct)$,
$\frac{1}{2}c^{-1}(\alpha - x + ct)\ (\alpha - ct < x < \alpha + ct < \alpha + x)$,
$t(ct < x < \alpha - ct)$, $\frac{1}{2}c^{-1}\alpha(\alpha - ct < x < ct)$,
$\frac{1}{2}c^{-1}(x + ct)\ (x < \alpha - ct < \alpha + x < \alpha + ct)$.

2.9 (i) When $x < ct$,

$$u = \frac{a(ct-x)+a(ct+x)}{2} + \frac{1}{2c}\left(\int_0^{ct-x} + \int_0^{ct+x}\right) b(s)\mathrm{d}s.$$

When $x > ct$, u is the d'Alembert solution.

(ii) When $x < ct$,

$$u = \frac{\alpha a(ct-x)+a(ct+x)}{2} + \frac{1}{2c}\left(\alpha\int_0^{ct-x} + \int_0^{ct+x}\right) b(s)\mathrm{d}s,$$

where $\alpha = (T - Rc)\,(T + Rc)^{-1}$.
When $x > ct$, u is the d'Alembert solution.

2.10 $g = 0\ (\zeta < h),\ 2e^{-k(\zeta-h)} - 1\ (h < \zeta < 2h),$
$-2(e^{kh} - 1)e^{-k(\zeta-h)}\ (\zeta > 2h).$

2.13 $-ck\,(2 + ck)^{-1} f(-x - ct),\ 2(2 + ck)^{-1} f(x - ct)$

2.14 The direct reflection from O is followed by a series of waves which have been reflected at L. The second of these waves is
$4c_1c_2(c_2 - c_1)\,(c_1 + c_2)^{-3} f(-x - c_1t + 4Lc_1c_2^{-1})$.

2.16 $-3\epsilon L^3/32$.

2.17 $4L/c$.

2.18 $u = \alpha \sin(x + t) - \frac{1}{6}xt^3$.

2.19 $u = \frac{1}{2}\tan^{-1}(x + t) - \frac{1}{2}\tan^{-1}(x - t) + t - \sin t + 2\sin x - 2\sin x \cos t$.

2.20 $u = a\cos(x + t) - \frac{1}{2}x^2t^2$.

3.3 $u = a \sin\dfrac{\pi x}{L}\cos\dfrac{\pi ct}{L} + \dfrac{a}{2}\sin\dfrac{2\pi x}{L}\cos\dfrac{2\pi ct}{L}$.

3.5 $u = \dfrac{8\lambda L}{\pi^2}\sum_{n=0}^{\infty}\dfrac{(-1)^n}{(2n+1)^2}\sin\dfrac{(2n+1)\pi x}{2L}\cos\dfrac{(2n+1)\pi ct}{2L}$.

3.6 Take $x = 0$ at the mid-point; the solution must then be an odd function of x. Let $u_t(x, 0) = (2\rho\epsilon)^{-1}I$ in $\frac{1}{2}L - \epsilon < x < \frac{1}{2}L + \epsilon$ and zero otherwise, and then take the limit of the result as $\epsilon \to 0$.

$$u = \frac{2I}{\pi\rho c}\sum_{n=1}^{\infty}\frac{1}{n}\sin\frac{n\pi}{3}\sin\frac{2n\pi x}{3L}\sin\frac{2n\pi ct}{3L}, \quad \text{where } c = \sqrt{(T/\rho)}.$$

3.7 $u = \dfrac{4L^2}{\pi} \sum_{n=1}^{\infty} \dfrac{(-1)^n}{(2n+1)\,[(n+\frac{1}{2})^2\pi^2 - L^2]} \left\{ \sin t - \dfrac{L}{(n+\frac{1}{2})\pi} \sin(n+\tfrac{1}{2})\dfrac{\pi t}{L} \right\}$

$\times \cos(\pi + \tfrac{1}{2})\dfrac{\pi x}{L}$.

3.12 $a_n = c_n + c_{-n},\ b_n = i(c_n - c_{-n})$.

4.1 1; $\cos nx$, $\sin nx$.

4.2 First find normal solutions of the form $X(x)e^{-i\omega t}$, where $X = A \sin k_1(x + L)\ (x < 0)$ and $X = B \sin k_2(x - L)\ (x > 0)$.

4.5 Normal modes
$Y_0[\lambda\sqrt{(L+h)}]J_0[\lambda\sqrt{(x+h)}] - J_0[\lambda\sqrt{(L+h)}]\,Y_0[\lambda\sqrt{(x+h)}]$,
where $4\omega^2 = \lambda^2 g$ and λ is any root of
$Y_0[\lambda\sqrt{(L+h)}]\,[\lambda\sqrt{h}J_0(\lambda\sqrt{h}) - 2J_1(\lambda\sqrt{h})]$
$= J_0[\lambda\sqrt{(L+h)}]\,[\lambda\sqrt{h}\,Y_0(\lambda\sqrt{h}) - 2Y_1(\lambda\sqrt{h})]$.

4.6 Replace the independent variable x by a new variable $y = \ln(1 + kx)$.

4.8 $\omega_n = \mu_n\sqrt{(g/3L)}$.

4.9 $p^2(2l - x)^2U_1'' + \omega^2U_1 = 0;\ U_2 = \pm A(2 + l^{-1}x)^{1/2}\sin[\lambda\ln(2 + l^{-1}x)]$.

5.4 $\dfrac{\omega^2}{c^2q^2} = \dfrac{\beta^2 + c^2q^2}{\lambda^2 + c^2q^2}$. $\dfrac{\partial \mathcal{E}}{\partial t} + \dfrac{\partial \mathcal{F}}{\partial x} = -2\rho\,\lambda u_t^2$.

$W = \dfrac{\bar{\mathcal{F}}}{\bar{\mathcal{E}}}$; averaging over a period of oscillation

$\bar{\mathcal{E}} = \tfrac{1}{2}\rho e^{-2px}(\beta^2 + c^2q^2),\ \bar{\mathcal{F}} = \tfrac{1}{2}\rho c^2 e^{-2px} q\omega$.

5.5 In a homogeneous material ω is a fixed function of k; $\omega(x, t) = F[k(x, t)]$.

5.6 $\theta =$ (i) $\tfrac{1}{4}\gamma^{-1}x^2t^{-1}$ (ii) $-\tfrac{1}{4}gt^2x^{-1}$ (iii) $-c^{-1}\beta\sqrt{(c^2t^2 - x^2)}$.
Phase lines are (i) parabolas, axis $x = 0$ (ii) parabolas, axis $t = 0$ (iii) hyperbolas.

5.7 10λ.

5.10 $u_{tt} = t^{-9/2}(\tfrac{3}{4}t^2 + 3itx^2 - x^4)e^{ix^2/t}$. $C(k) = \tfrac{1}{2}\pi^{-1/2}e^{i\pi/4}$.

5.11 $u = \dfrac{1}{\pi}\displaystyle\int_0^{\infty} \dfrac{\cos\frac{1}{2}k\pi}{1 - k^2} \cos kx \cos[(\beta^2 + c^2k^2)^{1/2}t]\,dk$.

6.3 $4m\omega^2\beta = (\alpha + 4\beta)^2 - (\alpha + 4\beta \cos ka)^2$.

7.2 $B = -\dfrac{P_{13}}{P_{11}}$, $C = -\begin{vmatrix} P_{11} & P_{12} \\ P_{13} & P_{23} \end{vmatrix} \Big/ \begin{vmatrix} P_{11} & P_{12} \\ P_{12} & P_{22} \end{vmatrix}$.

7.3 $u = \sum_{m=1}^{\infty} \sum_{n=1}^{\infty} (C_{mn} \cos\omega_{mn}t + D_{mn} \sin\omega_{mn}t) \sin\dfrac{m\pi x}{p} \sin\dfrac{n\pi y}{q}$,

where $\omega_{mn}^2 = \pi^2 c^2 (m^2 p^{-2} + n^2 q^{-2})$.
(a) $C_{11} = \alpha$; all other C_{mn} and D_{mn} zero.
(b) $C_{11} = \alpha$, $D_{22} = \beta/\omega_{22}$; all other C_{mn} and D_{mn} zero.
(c) $C_{mn} = 16p^2 q^2 (\pi^2 mn)^{-3} (1 - \cos m\pi)(1 - \cos n\pi)$; all D_{mn} zero.

7.4 $C_{mn} = 0, D_{mn} = 4Vp\,(cmn\pi^3)^{-1}(m^2 + n^2)^{-1/2}(1 - \cos m\pi)(1 - \cos n\pi)$.

7.6 $C_{mn} = 0, D_{mn} = \dfrac{16Vp}{cmn\pi^3\sqrt{(m^2+n^2)}} \sin\dfrac{m\pi}{2} \sin\dfrac{n\pi}{2} \sin\dfrac{m\pi\epsilon}{p} \sin\dfrac{n\pi\epsilon}{p}$.

$D_{mn} = \dfrac{4I}{\rho c\pi p\sqrt{(m^2+n^2)}} \sin\dfrac{m\pi}{2} \sin\dfrac{n\pi}{2}$, where ρ is the superficial density.

7.7 $c\pi p^{-1}\sqrt{13}$.

7.8 Following section 7.5 let

$$u = \sum_{m=1}^{\infty} B_m J_0(k_m r) \sin ck_m t, \text{ where } k_m p = \zeta_{0m}.$$

Then use (13.18) and (13.10). $B_m = \dfrac{2Vp}{c\zeta_{0m}^2 J_1(\zeta_{0m})}$.

7.9 $u = \sum_{m=1}^{\infty} A_m J_0(k_m r) \cos ck_m t$, where $k_m p = \zeta_{0m}$.

$$A_m = \frac{2\alpha}{p^2 J_1^2(k_m p)} \int_0^p (p^2 r - r^3) J_0(k_m r) dr = \frac{4\alpha p^2}{\zeta_{0m}^3 J_1(\zeta_{0m})}.$$

9.1 Given $\mathbf{E}(\mathbf{r}, t)$, define

$$\mathbf{H}(\mathbf{r}, t) = -\mu^{-1} \int_0^t \text{curl}\, \mathbf{E}(\mathbf{r}, s) ds + \text{curl}\, \mathbf{C}(\mathbf{r}),$$

where $\mathbf{C}$ is any solution of curl curl $\mathbf{C} = \epsilon\dot{\mathbf{E}}(\mathbf{r}, 0) - \mathbf{j}(\mathbf{r}, 0)$.
$\nabla^2 \mathbf{H} - \mu\epsilon\ddot{\mathbf{H}} = -\text{curl}\, \mathbf{j}$.

9.3 $\tau\{(np+1)^2-(n^2-1)(p^2-1)\sin^2 nkh\}=4np$;
(i) $h=\frac{1}{2}N\lambda$, (ii) $h=(\frac{1}{2}N+\frac{1}{4})\lambda$ (N an integer).

9.6 $|R^{\perp}/R^{\text{I}}|=1$ gives $|I^{\perp}/I^{\text{I}}|=1$.
$|\tan\frac{1}{2}\arg(R^{\perp}/R^{\text{I}})|=\frac{1}{4}\pi$ gives $n^2=3-\sqrt{2}-(\sqrt{2}\cos\theta-\sec\theta)^2$.

9.7 Energy $\frac{1}{2}\epsilon k^2(k^2-\beta^2)\int_D |F_3|^2\,\mathrm{d}S$. Dispersion relation is $k^2-\beta^2=$ const.
Group velocity $\mathrm{d}\omega/\mathrm{d}\beta=c\mathrm{d}k/\mathrm{d}\beta=c\beta/k=\beta\omega/k^2$.

9.8 On the inner conductor the current density is longitudinal and axially symmetric; total current is $2\pi Ac^{-1}\mu^{-1}e^{ik(z-ct)}\hat{\mathbf{z}}$. An equal and opposite current flows in the outer conductor.

10.6 The steady velocity field cannot be represented by a velocity potential because it is not irrotational. Write $\mathbf{q}=u(1+h^{-1}y)\mathbf{i}+\nabla\phi$, and show that ϕ must satisfy (10.11) and (10.13). Then replace (10.12) and (10.14) by relations derived from $\mathbf{q}.\nabla(\eta-y)+\eta_t=0$ on the surface and $\mathbf{q}_t+\mathbf{q}.\nabla\mathbf{q}+\nabla(\rho^{-1}p+W)=0$. After suitable approximation the conditions at $y=0$ are found to be
$\phi_y=u\eta_x+\eta_t$ and $\phi_{tx}+u\phi_{xx}+h^{-1}u\phi_y=-g\eta_x$.

10.7 Period of nth mode is $2\sqrt{(\pi L/ng)}$.

10.8 First show that if the width of the canal is $f(x)$ the surface profile satisfies the equation $f\eta_{tt}=g(hf\eta_x)_x$. In this problem h is constant and $f(x)=Cx$ (C constant). Assume $\eta=X(x)\cos\omega t$. The result is
$J_0(\lambda L)\eta=J_0(\lambda x)b\cos\omega t$, where $\lambda=\omega/\sqrt{(gh)}$.

11.2 (a) $u=xt\,f(x^{-1}-t^{-1})$.
(b) Let $u=(x+t)v$. $u=-x-t+(x+t)^{-1}f(xt)$.
(c) $u=t^2+tf(t^{-1}e^{x/t})$.
(d) $u=-t^2-x^2+f(xt)$.

11.3 (a) $x\mathrm{d}x+3t\mathrm{d}t-u\mathrm{d}u=0$, $\mathrm{d}x+2t\mathrm{d}t-2\mathrm{d}u=0$;
$c_1=x^2+3t^2-u^2$, $c_2=x+t^2-2u$;
$c_1=2c_2$ on Γ; $x^2+t^2-2x=u^2-4u$.

(b) $\mathrm{d}x+\mathrm{d}t+\mathrm{d}u=0$, $u\mathrm{d}(x-t+u)=2(x-t+u)\,\mathrm{d}u$;
$c_1=x+t+u$, $c_2=(x-t+u)u^{-2}$;
$(c_1-1)^2+(c_2-1)^2=2$ on Γ,
$(x+t+u)^2u^4+(x-t+u)^2-2(x+t+u)u^4-2(x-t-u)u^2=0$.

(c) $\cos(x-t)$.

(d) $c_1 = x^2 + t^2 - 2u$, $c_2 = xtu$; $c_1 + 2c_2 + 2 = 0$ on Γ;
$2u(1-xt) = x^2 + t^2 + 2$.

(e) $c_1 = u - \ln(x+t)$, $c_2 = x^2 - t^2$;
$u = \ln(x+t) + \sqrt{(\frac{4}{3}x^2 - \frac{4}{3}t^2)} - \frac{1}{2}\ln(3x^2 - 3t^2)$.

11.4 As the choice of new variables is not unique, canonical forms other than those given below are possible.
(a) Elliptic. $\phi = x^2$, $\psi = y^2$; $U_{\phi\phi} + U_{\psi\psi} = -\frac{1}{2}(\phi^{-1}U_\phi + \psi^{-1}U_\psi)$.
(b) Hyperbolic. $\xi = xy$, $\eta = y/x$; $U_{\xi\eta} = \frac{1}{2}\xi^{-1}U_\eta$.
(c) Parabolic. $\xi = y/x$, $\eta = x$; $U_{\eta\eta} = 0$.
(d) Parabolic. $\xi = x - y$, $\eta = x$; $U_{\eta\eta} = -U_\xi - 2U_\eta$.
(e) Elliptic. $\phi = y - x$, $\psi = 2x$: $U_{\phi\phi} + U_{\psi\psi} = \frac{3}{4}U_\phi - \frac{3}{2}U_\psi - \frac{1}{4}U$.
(f) Hyperbolic. $\xi = x - y^{-2}$, $\eta = x + y^{-2}$. $U_{\xi\eta} = 0$.

11.6 Let $\xi = \frac{1}{2}(t+x)$, $\eta = \frac{1}{2}(t-x)$; then $u_{\xi\eta} = 0$,
where $u_\xi = A(\xi)$. On the arc Γ, $\sqrt{\xi} + \sqrt{\eta} = 1$; $u = 0$,
so $u_\xi + u_\eta\,(\mathrm{d}\eta/\mathrm{d}\xi) = 0$, giving $\sqrt{\eta}u_\xi = \sqrt{\xi}u_\eta$,
and $1 = u_t = \frac{1}{2}u_\xi + \frac{1}{2}u_\eta$. Hence $A(\xi) = 2(1 - \sqrt{\xi})$,
and $u = -\frac{2}{3} + 2t - \frac{1}{3}\sqrt{2}(t+x)^{3/2} - \frac{1}{3}\sqrt{2}(t-x)^{3/2}$.

11.7 $-\frac{1}{6}xt^3$.

11.9 Riemann function v satisfies $[\eta(\xi v)_\xi]_\eta = 0$, giving $v = \eta^{-1}A(\xi) + \xi^{-1}B(\eta)$.
Boundary conditions on v then give $v = \xi_0\eta_0\xi^{-1}\eta^{-1}$, whence
$u = 2\xi\eta^2 - \xi^2\eta$.

11.11 Let $\xi = xt$, $\eta = x/t$. Then $2\xi u_{\xi\eta} = u_\eta$,
giving $u = \sqrt{\xi}A(\eta) + B(\xi) = \sqrt{(xt)}\,A(x/t) + B(x\,t)$.
Boundary conditions then give
$u = \frac{2}{3}x^2t^{-1} + \frac{1}{3}x^2t^2 - \frac{1}{5}x^3t^{-2} + \frac{1}{5}x^3t^3$.

11.12
$$u = \frac{a(x-t) + a(x+t)}{2} + \frac{t}{2}\int_{x-t}^{x+t} a(s)\frac{iJ_1(ip)}{p}\,\mathrm{d}s + \frac{1}{2}\int_{x-t}^{x+t} b(s)J_0(p)\,\mathrm{d}s,$$
where $p = \sqrt{[t^2 - (s-x)^2]}$.

11.13 $(\xi + \eta)^3 u = 2\xi^6 + 3\xi^5\eta - 3\xi\eta^5 - 2\eta^6$.

12.1 We are concerned only with $t \geqslant 0, r \geqslant 0$.
For $|r-ct|>R$, $u=0$. For $r+ct>R>|r-ct|$;
(a) $u=\frac{1}{4}c^{-1}\alpha r^{-1}[R^2-(r-ct)^2]$, (b) $u=\frac{1}{2}\beta(1-ctr^{-1})$.
For $r+ct<R$; (a) $u=\alpha t$, (b) $u=\beta$.

12.2 The answer is given by the Poisson solution with the initial conditions extended throughout all space by extending the functions a and b as odd functions with respect to z for $u(x, y, 0, t) = 0$, or as even functions for $u_z(x, y, 0, t) = 0$.

12.3 For $|r-ct|<R$, $u=0$; for $r+ct>R>|r-ct|$,
$u=\frac{1}{2}(1-ct\,r^{-1})\cos[\frac{1}{2}\pi R^{-1}(r-ct)]$; for $r+ct<R$,
$u=\frac{1}{2}(1+ct\,r^{-1})\cos[\frac{1}{2}\pi R^{-1}(r+ct)]+\frac{1}{2}(1-ct\,r^{-1})\cos[\frac{1}{2}\pi R^{-1}(r-ct)]$.

References and Additional Reading

Bouwkamp, C. J., *Reports on Progress in Physics* **17**, 35-100, 1954.

Coddington, E. A. and Levinson, N., *Theory of ordinary differential equations*, McGraw Hill, 1955.

Coulson, C. A. and Jeffrey, A., *Waves*, Longman, 1977.

Courant, R. and Hilbert, D., *Methods of mathematical physics*, Interscience, Vol. I 1953, Vol. II 1962.

Crawford, F. S., *Waves*, Berkeley physics course Vol. 3, McGraw Hill, 1968.

Goult, R. F., *Applied linear algebra*, Ellis Horwood, 1978.

Jackson, J. D., *Classical electrodynamics*, Wiley, 1975.

Jeffreys, H. and B. S., *Methods of mathematical physics*, Cambridge University Press, 3rd ed. 1956.

Landau, L. D. and Lifshitz, E. M., *Theory of elasticity*, Pergamon, 1959.

Lighthill, M. J., *Waves in fluids*, Cambridge University Press, 1978.

Morse, P. M., *Vibration and sound*, McGraw Hill, 1948.

Panofsky, W. K. H. and Phillips, M., *Classical electricity and magnetism*, Addison-Wesley, 2nd ed. 1962.

Rayleigh, Lord, *Theory of sound* Vol. II, Macmillan, 1896.

Sneddon, I. N., *Elements of partial differential equations*, McGraw Hill, 1957.

Sommerfeld, A., *Optics*, Academic Press, 1954.

Tolstov, G. P., *Fourier series*, Prentice-Hall, 1962.

Whitham, G. B., *Linear and non-linear waves*, Wiley, 1974.

Index